# The Avid Handbook

**Steve Bayes**

**Focal Press**
Boston • Oxford • Johannesburg • Melbourne • New Delhi • Singapore

Focal Press is an imprint of Butterworth–Heinemann.

Copyright © 1998 by Butterworth–Heinemann

 A member of the Reed Elsevier group

All rights reserved.

Avid and the Avid logo are registered trademarks and AvidPress is a trademark of Avid Technology, Inc.
All other trademarks found herein are property of their respective owners.

No part of this publication may be reproduced, stored in a retrieval system, or transmitted in any form or by any means, electronic, mechanical, photocopying, recording, or otherwise, without the prior written permission of the publisher.

∞ Recognizing the importance of preserving what has been written, Butterworth–Heinemann prints its books on acid-free paper whenever possible.

 Butterworth–Heinemann supports the efforts of American Forests and the Global ReLeaf program in its campaign for the betterment of trees, forests, and our environment.

**Library of Congress Cataloging-in-Publication Data**
Bayes, Steve, 1959-
    The Avid handbook / Steve Bayes.
       p.   cm.
    Includes index.
    ISBN 0-240-80347-7 (paperback : alk. paper)
      1. Video tapes—Editing—Data processing.   2. Motion pictures—Editing—Data processing.   3. Media composer.   4. Avid Xpress.
I. Title.
TR899.B37    1998
778.59'3—DC21                                      98-2566
                                                            CIP

**British Library Cataloguing-in-Publication Data**
A catalogue record for this book is available from the British Library.

The publisher offers special discounts on bulk orders of this book.
For information, please contact:
Manager of Special Sales
Butterworth-Heinemann
225 Wildwood Avenue
Woburn, MA 01801-2041
Tel: 781-904-2500
Fax: 781-904-2620

For information on all Focal Press publications available, contact our World Wide Web home page at: http://www.bh.com/focalpress

10 9 8 7 6 5 4 3 2 1

Printed in the United States of America

Dedication

To Mary Andress, Marjorie Bayes, and Jane MacFarlane for all the birthdays

**Thanks to:**

Peter Bos
Joe Doyle
Jamie Fowler
Mark Geffen
Dave Meichsner
Stephen Hullfish
Jim McKenna
Don Nelsen
Paul Pearman
Greg Staten
Tim Vandawalker

*"It is circumstance and proper timing that give an action its character and make it either good or bad."*

— Agesilaus, 444-400 BC

*"A globe-full of people, and not one is ignorant of the worth of twenty minutes, each minute a Pearl, let slip, one after the next, into Oblivion's Gulfs."*

— Thomas Pynchon

# Table of Contents

| | | |
|---|---|---|
| **Preface** | | xi |
| **Chapter 1** | Talking to the Machine | 1 |
| **Chapter 2** | Workflow of a Nonlinear Project | 5 |
| **Chapter 3** | Intermediate Techniques | 27 |
| **Chapter 4** | Administration of the Avid | 47 |
| **Chapter 5** | Long Format and Improving System Performance | 77 |
| **Chapter 6** | Importing and Exporting | 85 |
| **Chapter 7** | Preparing for Linear Online | 105 |
| **Chapter 8** | Finishing on the Avid | 127 |
| **Chapter 9** | Simple Troubleshooting | 163 |
| **Chapter 10** | Introduction to Effects | 185 |
| **Chapter 11** | Nonlinear Video Assistants | 201 |
| **Index** | | 209 |

# *Preface*

*The Avid Handbook* is a book about nonlinear editing that can be read in a linear fashion, but it can also be kept handy to refresh your memory before you start a big job. It will give you the confidence you need after taking the introductory Avid course to deal with the more complex situations that occur outside the classroom.

This is not a tips-and-techniques type of book; there are already several of those and they must be updated every year as new keystrokes are added and other keystrokes are eliminated. A series of keystrokes will only help you under very specific circumstances and does not really represent a coherent view of the machine or its capabilities. Many people who are looking for a tips book are striving for a deeper understanding of ways to make things go faster and smoother and give them an edge over their competition. You won't get that by memorizing even more arcane variations on secret handshakes. You will gain the power of the technology as part of new production *procedures*, which is the way this book is focused.

Other people will want a beginner's book of "How to Turn It On and Start Making Money" but this type of book devalues the depth and flexibility of this top-of-the-line tool. For better or worse, Avid editing systems are complex because they deal with complex, ever-changing, and unpredictable situations. Take a class or find a good teacher and spend the time with an experienced human to get going and learn the basics. Having a good teacher is clearly worth the money. It's your career—start this part right!

This book is geared more toward video than film for several reasons. First, there are already several comprehensive books about film and the technology. Second, this technology has completely changed the production methods of the video world and the impact has not yet been completely explored. The completely digital video production is a reality today, but many years off for the film community and as a result, the procedures in video have changed more radically. Still, much of the material in this book applies to both video and film. Even if you see methods that cannot be applied to your film situation, there may be a spark of recognition as you take several of these techniques and combine them or take the spirit of an idea and modify it.

This book is really intended for overworked editors, assistants, or post-producers who finds themselves needing to know more than they really have any chance of learning through their limited experience. They are at a stage bordering on intermediate, but are being asked to make predictions and estimates based on procedures they have never seen and functions they have never used. Everyone in the industry should have a healthy respect for the vast amount of things that can go wrong and the increasing levels of complexity in *any* production. A complete overview of procedures will show you everyday savings, shortcuts, and strategies. Display *The Avid Handbook* proudly in your Avid suite and reread it while you are digitizing and rendering!

# 1

# *Talking to the Machine*

For artists and craftspeople working today in the communications and entertainment industries, the main characteristic that will distinguish this time period from others will be the speed of change. Tools have always evolved and clever workmen have always adapted their existing tools to the job at hand. Once a tool has been refined to a certain point, it may stay in that particular state of grace for centuries. No such luck these days.

Coming to grips with new techniques and features is a constant and never-ending process. You now face an everyday choice of whether to exchange the power of the newest and the fastest for the comfort of the familiar.

But maybe it's not really so bad. Maybe there is a thrill to the new and different, a pleasure in finding the best way to work for you. The secret pleasure of a complex system is simplifying the tools to meet your needs, streamlining them so that you are in complete control. You don't need to buy new software for a new job, just get out the manual and find out what is hidden inside this one giant piece of software. It's a treasure hunt to find the extra bit of power that can propel you through a tedious task and finish while your client is still on the phone. Or like a magic trick, you do have a few things up your sleeve, capabilities your clients have never seen before, things you have developed for yourself that can astonish and amaze.

That's the positive side to all this change. You don't need to know everything. There is no way you can know everything. The time you spend memorizing arcane computer jargon and seldom-used techniques is time away from the creative process and time away from friends and family.

This book is meant to be the next step after formal training. I strongly recommend taking a formal beginner's course. I helped write the courseware (along with a large staff of experienced technical writers and teachers) and I certify the instructors. No matter how experienced you are, if you do not start with formal training you will have odd gaps in your knowledge. You are forced to guess and improvise when someone has already thought it out for you. After viewing a demo reel filled with beautiful, multi-layered graphics, I asked the editor to send me his sequence on a floppy. "How do you do that?" he asked. An odd gap, but one that is holding him back.

This handbook is meant to fill in those gaps. It provides the extra level of depth that you can't get in a class full of techniques; the type of experience and insight that comes from watching beginners screw up and advanced people fly; the type of experience that takes a long time to sink in. Along with this handbook you should experiment, read the user documentation, take an advanced course, read the Internet, and never stop learning.

By necessity, training for adults must be short and intense. Unfortunately, it is also hard to retain all the information taught in a short course. When I train trainers how to teach the Media Composer, my mantra is "Learn what to leave out." In two, three, or five days you can do more damage trying to overload your students. Make them feel confident and competent with the basics. You will not be "an Avid Editor" after a short course, but you will be able to start and complete your first job. Sometimes I feel an evil joy in telling my students that I am always learning something new on the Avid, after eight years. They don't think that's funny.

A very important translation goes on when you take a vague idea in our heads and translate it into pull-down menus. I have found a little trick that allows the creative to cohabitate with the technical side of my brain. The trick stems from the understanding that the computer is only doing what you ask it to do. The computer is interpreting those commands through the filter of the minds of the people who programmed it. If the designers of editing software don't understand professional post-production you will have to fight and trick the machine at almost every step beyond the most basic tasks.

When you sit in front of an Avid interface you trust that the Avid designers understand what you want to do and, if they have not written something to accomplish exactly that, at least they haven't held you back with arbitrary restrictions. You have some freedom to improvise and many of the techniques described in this book have been developed because the system is so flexible.

I've taught Avid editing around the world for the last four years. What was really interesting was that I was learning as much as my students. I learned about sitcoms and army VIP videos, promotion departments and daily magazine news shows. I saw this same piece of software used every day in many completely different ways. I understood very clearly that there was no right way to do most things and what is a minor problem for some is a catastrophe for others. Features that you have never seen anyone use will regularly save entire productions. I had to throw away a lot of preconceptions and realize that the more I learned the less I knew.

The value for me in teaching Avid in so many different environments is that I get to experience different needs and ways of doing things. The purpose of this handbook is to synthesize all of these experiences into an easy form for you to apply immediately. While using the system for the last eight years I have seen hundreds of situations where just a little advantage is enough to keep a client or make the difference between losing money, breaking even, or making a profit.

This handbook describes procedures that go beyond the basic functions of the system and adapt to changing situations and demands. Knowing that the procedures are available will allow you to plan more efficient strategies. The more you understand the capabilities for organizing and the database power of Media Composer and Avid Xpress Mac, the more you will be able to plan for and stay calm during the inevitable chaos.

So what should you focus on to learn and master the Avid editing systems, Media Composer and Avid Xpress? Although each system has a different market, there are some very important similarities both in the software and the way they are used. This handbook does not focus on the individual menus or buttons, although they must be covered to get any substantial hands-on value. Buttons and menus change and go out of date so fast that standard printed books no longer can really keep up when you get too specific with professional software. But the ideas and the procedures in this book should help you for some time to come because they deal with the entire process and where the Avid fits in.

*The Avid Handbook* is being researched and written while Avid continues to develop the next version of the software which will be version 7.0 for Media Composer and version 2.0 for Avid Xpress. The software will have shipped by the time this book is available so the new capabilities are being included while the software is still being beta tested. Promising anything about unreleased software is not very safe, but having worked on alpha and beta versions for eight months I am convinced that including some mention of it in this handbook is valuable. Some of these techniques may not work exactly the same way with the version of the software that finally ships, but since this handbook focuses on situations that will not change much, the general ideas and procedures will still be valid even if the pull-down menus get moved around. Unless there are giant leaps in technology, these methods or similar ones, with your own personal variations, should still be in practice.

There will be times when a particular function does not exist in Avid Xpress that is standard in Media Composer, but for the most part a technique described in this handbook will make sense anyway and may give you some new ideas about how to use Avid Xpress. Most of the time I will refer to the Avid software as the "MC/AX" to show that both are included. Even if you are only using your Avid Xpress once a week, it is still worthwhile to learn these techniques and polish your skills to be more proficient. There is very little that deals specifically with NewsCutter and the newsroom because it is a unique environment. There should be some good material in here for the newsroom, but most of these procedures favor highest quality over quickest method when you are forced to make a choice, although both may be mentioned. Hopefully, you are not forced to make that choice too often!

This handbook makes some assumptions about basic understanding of the Macintosh; otherwise this book would take on the proportions of combined Avid and Macintosh documentation. This handbook will replace neither! There will be some complex techniques to help you deal with complex situations. I like to think of them as recipes, the kind of lists of instructions you deal with every day. I have tried to include the easy techniques, too. If you find yourself a little confused, use the simple way although, many times, it will not be the fastest.

I hope this handbook helps you to avoid down time and maximizes your creative time. Get past the bells and whistles to the really important stuff and have confidence that with better tools you can become a better editor. You're working with the best system in the world for telling stories with pictures and sound. You will never go back to working the old-fashioned way. In fact, you could forget how tedious it used to be, quickly adapt, and take it all for granted.

# 2

# Workflow of a Nonlinear Project

Workflow is where many beginning editors and producers can take advantage of new opportunities and improve their efficiency by looking at the whole process. The main steps to any nonlinear digital project are input, editing, and output. This chapter describes in depth strategies for these three basic steps and how to adapt to fit them easily into your production methods. You must get a clear picture of how this technology will change your scheduling and budget.

## CHANGE YOUR MIND WITHOUT LOSING IT

First let's deal with the most commonly repeated myth of the benefits of this technology, "It's faster." Well yes, the computer does arithmetic better than you, but it can't decide between green or blue for a title or whether a logo is too small. You still need to make all the decisions. The computer makes it easier and faster to implement those decisions and to see the many subsequent versions for the client's new vice president, but lets face it, humans don't make decisions faster because they are using a computer. In fact, some things take longer using the computer. Once you have a clear idea of where you want to go, the mechanics of getting there are accelerated.

You would have prudently stopped at the third or fourth version of a sequence using more traditional methods just because of the sheer exhaustion of creating them all or the loudly ticking clock of the expensive online suite. Now you are forced to defend your cut for ethereal reasons of timing, sensibility, and art, reasons that can take much longer to justify. It is also easier to lose that argument since, "The client is always right" or at least they are paying the bills. The freedom to experiment and explore can be a liability unless you understand how to take control of a situation, fight for what you think is the best, and know when you are just making the scene worse.

Multiple versions mean multiple screenings for various groups of people with conflicting and passionate opinions. Lots of discussion ensues with arm-twisting, insults, shameless self-promotion, and occasionally a decision. This is progress, but it's not fast.

Another reason that your job may not go faster is that since all of these changes are supposed to be so easy and painless, you will get them at the last possible moment. Before, there was a considerable penalty for changing your mind at the last minute and this put pressure on producers to be better prepared before the final stage of the project. It was easier to charge for the overages because everyone understood how hard those changes would be.

Some changes are easy with a nonlinear system. Making Part 3 into Part 2 takes just a few minutes, but the difference between an easy fix and a much more complicated one may not be obvious. Some changes have always been difficult, for instance, changing layer three in the middle of a twenty-layer effect. Those familiar with traditional layering methods will actually be surprised at how easy it is now to make that kind of change. What they aren't prepared for is all the little changes that ripple throughout the entire show, all the new details that add up to extend a project in unexpected ways. In the workflow of a nonlinear project, there are places where the work will go amazingly fast and, occasionally, the technology and the added capabilities will add some requirements to slow you down.

## INPUT

Let's look at the first step: digitizing. If you are working with the Ikegami EditCam you can skip this step. With versions 7.0/2.0, the MC/AX is compatible with the EditCam media. You import the shots from the fieldpak and all your media is online. However, even if you are not digitizing, you still need to incorporate pre-production planning and project organization in a way that will influence the shooting and logging.

### Logging

Logging is your link between production and post-production so don't treat it lightly! If you have strategies in place to minimize hand entry and duplication of efforts to input data to the MC/AX, you will have instant savings. Logging and analyzing as much of the footage as possible before post-production really begins is one of the best ways to improve the speed and efficiency of digitizing. If you are working on a film-based project, much of the initial logging will be taken care of for you during the film-to-tape transfer process. The transfer house will create a database, like a Flexfile, that can be imported as a shotlog into the Film Composer and the digitizing can start as soon as the film is transferred to video. Video-based projects must spend more time preparing for the digitizing stage.

MediaLog, the logging software that ships with every system, creates MC/AX bins on a Macintosh without any special hardware. You use a deck control cable that came with MediaLog and connect it from the tape deck to the serial port of the computer, the connection used for an external modem or a printer. If you can control a video deck with the Macintosh while using MediaLog, you can enter start and end points on the fly for digitizing your clips. MediaLog will read the timecode that is on the address track of the tape. You can then bring the floppy disk with the logged project to the Avid suite and copy the bins to the MC/AX. The digitizing is automatically controlled by the computer, which asks for each tape as it is needed.

If you have an internal modem, then the modem serial port on a PowerBook or a Duo is not available, even if it is the only serial port. On a PowerBook computer, you need to turn off the internal modem before you can use that serial port, making the PowerBook a better choice over a Duo. There is a limited range of professional decks which can be controlled without extra deck control hardware (a VLAN or a VLX) so make sure the deck you want to use is on the Avid list of supported decks in the user documentation. Avid uses standard Sony protocol, RS-422, but strangely, many popular decks don't comply with this standard.

There are versions of MediaLog available for DOS and Windows, but the files need to be converted to Avid Log Exchange (ALE) before they can be imported as a shot log and read by the Macintosh-based editing system. This is a good way to work if you are only logging by eye and not trying to control a deck or if your portable computer is a PC.

The biggest danger with MediaLog, besides entering the numbers wrong, is logging on a version that is different from the version of MC/AX you are using for editing. MediaLog 6.x will have problems sending bins back to Media Composer 5.5. Forward is good, backwards is problematic. If you must work with mixed versions of the software (a bad idea), make sure that the version of MediaLog is earlier than the version of MC/AX. If you are having problems getting your bins into the MC/AX, you can always display the most important headings in the bin and export from MediaLog as a shot log. Importing is much more forgiving using a text file instead of an unrecognizable version of a bin.

## Timecode Breaks

One of the most important points to consider when logging is timecode breaks in the field tape. Videotape shot in the field will have breaks in the timecode whenever the deck is completely stopped, not just paused. However, if the crew is shooting a time-of-day (TOD) timecode, you will have a break in the timecode every time the deck is paused. The TOD timecode generator continues to increment when the deck is paused and, when the tape starts rolling again, you have a timecode that reflects the new time of day. TOD timecode (sometimes called freerun) is especially problematic with a logging program and specifically with older versions of the MC/AX. I do not recommend using TOD unless you are shooting an event where there is continuous action, long takes, and multiple cameras will be running (like multicam). TOD is most useful for a music or stage performance, but a disaster if there are a lot of quick shots. The new exception to this rule is a feature in versions 7.0/2.0 called "Use control track instead of timecode for preroll" in the new Digitize User Setting. This allows you to preroll over a break in the timecode during the digitizing if there is no break in the control track, but it still means you will need to be extremely careful while logging.

MC/AX generally creates a new master clip whenever it digitizes across timecode breaks. You can just pop a tape in, set the Digitize Setting to "digitize across timecode breaks" and walk away. The MC/AX senses every time there is a break in the timecode, makes a new master clip, rolls forward to come up to speed for a few seconds, and then starts digitizing again. If there is plenty of time between the field deck starting and the beginning of the action, this works out fine. However, chances are that at least some of the takes will not come up to speed completely before the action begins and the beginning of the action is not digitized.

Make sure to set the preroll for the deck to be as short as possible. This still may not be enough because the MC/AX keeps rolling for a few seconds after it finishes digitizing a master clip. If you chop off the beginning of the action or you really want to keep the slate, then you need to occasionally modify the master clip and redigitize. Even if the action is not chopped off, there will be very little useful dead space before the action. You need those pauses for trimming, dissolves, and, as a last resort, roomtone. Starting the master clip at the very beginning of the action is a bad habit. Don't make the editing decisions now!

Preroll and postroll are the keys to good logging. You must have a minimum preroll of several seconds after the timecode stabilizes and the field deck comes up to speed. If you are working with Betacam SP then you only need about three seconds, but you should count on five seconds. U-Matic struggles with anything less than four or five seconds. Don't log the beginning of a shot until this amount of time has elapsed after the field deck has started recording or the deck will rock back and forth during digitizing as it tries to find the proper preroll point — those three to five seconds before the logged inpoint. If the Avid can't find the proper preroll point it stops and gives you the error message about "failed to find preroll coincidence point." The expected preroll point isn't there.

You can change the preroll under the "Deck Preferences" setting so that the preroll is shorter. (The default preroll time changes depending on the type of deck you choose in versions 7.0/2.0.) If you set a very short preroll time, you run the risk of the first few frames of the digitized shot being unstable or worse, when you make an edit decision list (EDL), the decks will abort the edit in the online suite. The online editor will have to change the preroll time for that shot

**Figure 2.1** Deck Settings Dialog Box

which may stop the auto assembly process, prompting the question "Who logged this?"

If you absolutely must have the first few frames of a shot and you have a break in the control track, you will not be able to log this shot. You must digitize on the fly by letting the deck roll over the break and quickly clicking on the big red digitize button to start the digitizing without preroll. However, if you must make an EDL, you are better off dubbing this shot to another tape because you will never be able to assemble it in the linear online suite.

## Using Modify

The best choice you have for fixing a preroll problem is to modify the logged clip and change the inpoint. Highlight the clip and click on the start time in the bin. You can change that number to anything. Alternately, you can go under the Clip Menu and pull down Modify. Modify allows you to type in a specific number to increment or decrement the starting timecode of the master clip and allows you to modify all the shots at once in a particular bin. In the case of a preroll problem, either increment the timecode by enough to get the start point away from the control track or timecode break, or decrement the timecode to get a few extra frames from the head of the shot and take your chances on a short preroll.

Modify is a very useful command which will be used many times throughout this book. It has just been added to the Avid Xpress for version 2.0, but is available on all versions of the Media Composer. Be aware that when you modify an inpoint on a master clip, the duration stays the same—the outpoint will move by the same amount. You may have just chopped off the end of the shot! You must also change the outpoint if you want the duration to change when you modify an inpoint. The MC/AX does not allow you to change duration to a shot if it has already been completely digitized because it protects you from messing up every potential sequence this shot was used in. Any change to a master clip will ripple across your entire project and affect every sequence and every bin that contains that shot. In order to change duration you must first "undigitize" which is really an "unlink" (shift-control-relink) operation. This allows you to change the inpoint and the duration and unlinks from sequences. This book will discuss linking and unlink in more detail later.

These are pretty easy fixes compared to the postroll problem. If you log the outpoint of a particular clip after a break in timecode you will not discover the error until after you have almost finished digitizing the master clip. You may have to wait until the end of a five-minute clip before you get the message that there is a discontinuity in the timecode and the entire material so far must be discarded. Then you must modify the outpoint and determine where the timecode really ends. It pays to double-check any written log you are handed for these exact reasons. After you type in the hand-written field logs you should always "go to" both the ins and outs (using the "Go To" buttons) to check that the numbers exist and have enough preroll and not too much postroll.

There are some methods of logging in the field "on the fly," without stopping the deck, and they may save you some time at the expense of complete accuracy. There are ways to connect the camera or the deck directly to a computer and bang the ins and outs as they happen using infrared signals from the field deck. These methods work best in a controlled environment like a studio, but some people have been known to do it with special battery packs for every piece of equipment.

As you begin to use MediaLog more you will discover that there are a few shortcuts. For instance, you should use the mouse as little as possible. You can tab through the various entry fields and use the B key to actually enter the log (make sure the Caps Lock key is off). You can then type the name of the clip and use Command-7 to get back to the logging interface and tab through again, all without using the mouse! Since logging is inherently repetitive, even these small shortcuts will seem like lifesavers later on.

## Tape Wear and Dubbing

Some producers are concerned about tape wear and the possibility of creating dropouts or creasing the tapes by playing them too often. This is a serious concern if there is much logging to do. Dropouts occur when small parts of the tape's magnetic coating, the oxide, flake off and stick to the playback heads. This is more noticeable with the smaller formats like Hi-8 or DVC since a small amount of oxide on a small format makes a more visible dropout. Although improvements in tape stock are continuing constantly, "prosumer" formats do not find it cost effective to use the more expensive materials or use better quality control when manufacturing these tapes. Usually you will find yourself dubbing to a higher format like Betacam SP and then using those tapes throughout the production while the originals are carefully packed away for a worst-case scenario. If a Betacam SP dub of your Hi-8 is lost or damaged, you have to go back to the original field tapes and re-dub them. If there is timecode on the original tapes then carry it over to the dubs (called jam-sync). This way you can make another set of dubs to match what you have already digitized. If there is no timecode on the original tapes you may be forced to re-edit by eye if the dubs are damaged.

In situations where tape wear, a long post-production schedule, and cost is a factor, even with a high-end source tape, dub all of your tapes to another format like 3/4" U-Matic, S-VHS, or even Betacam SP. Dubbing from a higher format like Digital Betacam to S-VHS or Betacam SP saves you money by reducing the cost of renting an expensive Digital Betacam deck. It also allows you to transfer the timecode so you have a perfect match back to the original sources when you redigitize for the finishing stage (using the original DigiBeta masters and renting the DigiBeta deck for a day or so) or when you make an EDL for the online linear assembly. Evaluate your cost savings and safety factor by the amount of tapes shot for the project and the extra time for dubbing. The cost savings are quickly undone if there are dozens of tapes to dub and dubbing is a cost that must go out of house.

## Logging by Hand

To reduce costs and tape wear even further, if you must use the originals when you digitize, there is the choice of dubbing everything to regular VHS just for logging. If you put a timecode window burned in to either the upper or lower part of the video frame, you can use this as a visual reference and something you can easily show the client for comments. There is no actual timecode on this tape, just the display of timecode, so you can't actually use these dubs to digitize with. This requires the producer or the editor to view these tapes in the comfort of their own home (or some other place where they don't get charged by the hour) and control the tape with a standard remote control on a VHS deck that allows a good clear freeze frame. Unfortunately, this method requires that someone type by hand the timecode numbers for the start and end of each shot into MediaLog. This option is faster if you purchase the number keypad attachment that connects to your PowerBook's ADB port (where the mouse goes) and have a VHS deck with a search knob for finer control.

Some people have VHS decks that can read VITC, a timecode recorded in the video signal, and can be controlled by the Sony deck protocol, but this is rare.

One product, Easy Reader™ from Telcom Research makes logs based on timecode breaks when controlling a consumer deck if you have recorded VITC on the VHS tape when making the dubs.

There are some challenges with typing numbers in by hand that have to do with reducing human error. First, what kind of timecode has been recorded: drop-frame (DF) or non-drop-frame (NDF)? PAL people don't laugh, someday you will be forced to work in NTSC too! How can you tell the difference and, um, what is the difference?

## Drop-frame vs. Non-Drop-frame

Here is a quick primer on drop-frame and non-drop-frame. NTSC videotape runs at 29.97 frames per second, not 30 frames per second (fps). If you count frames at 30 fps then you have an interesting problem. A show which uses timecode that increments at 30 fps will actually have a running time longer than one hour even though the timecode says it is exactly 1:00:00:00 for a duration. Because of the slight difference between 29.97 fps (the true playback frame rate) and the more convenient 30 fps, at the end of one hour a non-drop-frame master is too long by exactly 108 frames. This is enough to cut off the credits or worse, interfere with the commercial coming next! The two frame rates do not change the speed of the tape, just the way that the frames are counted. In fact both drop-frame and non-drop-frame both run at 29.97 fps, but non-drop-frame counts the frames at 30 fps. In order to correct for the frame count not matching playback time, certain frame numbers are skipped in drop-frame timecode. Frames are not actually *dropped*, but two frame numbers per minute, except for the tenth minute, are not used. This means that when using drop-frame and the timecode says one hour, the realtime playback of the show will take exactly one hour and not one hour, 3 seconds, and 18 frames.

Since accurate timing is most important for broadcast purposes, drop-frame is traditionally used in broadcast. Non-drop-frame is used in corporate work, film transfers for cutting at 24 fps, and other nonbroadcast uses like laser discs. This is mostly because the math is easier to do on a standard calculator with non-drop-frame and all possible frames can be searched for by a computer interactive playback system like a kiosk, multi-screen display or games. There are timecode calculators however, and the Media Composer and Avid Xpress contain a pretty good one in the Tools menu and, really, once you choose one format or the other the choice is no longer apparent until it is time to output.

How does the computer know what kind of timecode is on your tapes? In versions 6.5/1.5 and later, when you pop a tape into a deck that is connected to the MC/AX, the computer plays a few seconds of that tape automatically. The software reads what kind of code is on the address track before it allows you to name a tape. In earlier versions you must play the tape yourself for a second in the deck before naming it.

**Figure 2.2** Deck Preferences Dialog Box

What if you are not controlling the deck with a computer? You must enter the type of timecode by hand. Look at the burn-in timecode in the small window display on the VHS dub. Look at the characters between the last two digits, the frames, and the next two digits, the seconds. If the character is a colon (:) you are working with non-drop-frame. If it is a semi colon (;) you are working with drop-frame. There is a Deck Preferences Setting which says, "If there is no tape in the deck log as" and then a choice between DF and NDF. Well, there is a tape in the deck! But, you are not controlling the deck (it is not physically connected to the computer), so this setting applies. Choose the setting that matches your burn-in timecode and make sure that this type of timecode does not change! If the timecode type in your timecode window dub changes, you have to go back to the Deck setting and change the type of timecode there.

There is a difference between drop-frame on the field tapes and drop-frame on the finished master. You can use any kind of timecode on the field tapes and you can mix and match as you create the finished master tape. The master tape will have one type of timecode throughout and the source tape does not influence this. It is a good idea to have the timecode type in your project be consistent, but sometimes this is not possible because of the use of historical or stock footage. Camera people who always shoot one type of timecode may be unwilling or unfamiliar with how to change the type of timecode recorded on their field deck. One freelance crew may shoot non-drop and the next may shoot drop. This isn't critical, just annoying. To be fair, on some field decks you need to pull out a circuit board to make the change. If you have mixed DF and NDF in your project you will need to be extremely vigilant during the logging process.

## The Importance of Tape Names

When logging in either MediaLog or MC/AX, start by creating a new tape in the project and give that tape a simple and unique tape name. Tape name is very important at this stage. Short names are good and a name consisting of all numbers is even better, especially if you are planning to create an edit decision list (EDL). When you create a new tape name you are assigning properties to this tape. Attached to the tape name is an indication whether this tape is drop-frame or non-drop-frame. (This is not an issue in PAL.) The second property is the name of the project this tape is in. Tape 001 in Project X is a completely different tape than Tape 001 in Project Y. Keep this in mind when you move from a project created in MediaLog and a project created in the MC/AX.

Keep all tapes in the same project in which they were logged. This means you shouldn't create a new project when you get to the digitizing stage, just copy over the project you started with in MediaLog. If you have logged on a PC, you will have an ALE (Avid Log Exchange) file that you import to the project on the MC/AX. It automatically becomes part of your project, but if you import it twice, the system thinks these are different takes on a different tape with the same name. This will have consequences for your EDL, so don't do it.

There is one more challenge with this method. When your tapes were dubbed, they possibly put four, 30-minute long field tapes onto a 2 hour VHS. When the field tape changes on the dub, treat this as if you had just changed tapes and create a new tape name. When the tape changes, pay attention to what kind of timecode this new tape is. The tape name you created when you logged the first field tape on the VHS will accurately track back to the original field tape.

If you do not change the tape name and log another field tape you will have problems digitizing. The computer tries to find Tape 001 at timecode 5:15:30:00 when the tape may have changed several times on the VHS dub and the real Tape 001 ended at 1:29:15:00. When digitizing, Tape 001 fast forwards to the end of the reel as the Avid searches for a timecode that is actually on the next tape or several tapes later (probably Tape 005). You will get an error message that the Avid "failed to find timecode coincidence point." You must then tediously find all the shots that start with a five-hour timecode and modify the source name to be Tape 005.

Consistency of timecode type and attention to details that allow you to track back from dubs to the originals is important to prevent confusion. The system tells you there is a problem when you have logged tape 001 as NDF and you pop in a DF tape and say "Sure, this is Tape 001" by clicking "OK" when it asks for Tape 001. Then you will get the following message: "The tape you have entered is drop-frame and has been logged as non-drop-frame or it is non-drop-frame and has been logged as drop-frame." This basically means that the field tape's timecode does not match the tape name that was logged. If this is really the right tape, then a mistake was made during the logging and, when the tape name was assigned, the system was either reading the wrong type of timecode or told to assume the wrong type of timecode because of the Deck Preferences Setting. You must modify clips so they are the right type of timecode and assign them all to a new tape name that has the same type of timecode.

## Creating the Database

You are creating the main database for this project whenever you create electronic logs. You can use this database to sort, sift, and search for any and all kinds of information later. Plan to have several custom headings on large jobs that will allow you to search across many bins to find an obscure shot. Although the MC/AX does not allow you to search across bins, there are third-party applications that will allow you to look inside even closed bins. Retrieve-It! from MVP Solutions is the best program I have seen for this because it allows you to peek inside the programming code of a closed bin without actually opening it. Since large bins take some time to open, this can really allow you to fly along. Keep in mind, though, that a third-party search engine like Retrieve-It! does not know to look for the 15 in the Scene column. Retrieve-It! doesn't know about columns. Be sure that any search criteria you create is valid without the column information, like "15/Lake view WS."

## Alternatives to MediaLog

You don't need to use MediaLog to log your project. You can use any word processing, spreadsheet, or database program as long as you format it correctly. There is a standard way to set up the headings and then tab-delimit the information according to the format set forth in the user documentation. You can then convert the file to a text-only document and import to MC/AX as a shot log. You can create a template in a program like FileMaker Pro and create your bin information to import to MC/AX to fit that template.

If you export from MC/AX as a text file that is tab delimited, then FileMaker Pro enters the column information into the right headings in your template. You can then search on 15 in the Scene field of the database template. This takes some setting up but can be very powerful.

MediaLog ships with every Avid editing system and is by far the simplest to use since it is almost identical to the digitizing interface of the MC/AX. Some people like other programs because with MediaLog you can't start entering information about a master clip until you mark an outpoint. You can fast forward to the end of a clip and mark it and then enter all the heading information. But if you want to view the tape in real time and enter the heading information at the same time, then you might want to consider a database or a spreadsheet program. There are also programs on the market, like Executive Producer by Imagine Products, which create complex databases using the Avid text format with screen grabs and multiple ways to format the information for presentation purposes.

## Organizing Bins

An editing system so heavily dependent on the computer means you must spend more time getting organized at the beginning of the job. It may appear like you are off to a slow start but again, when you get to the fast part, the editing, you will look like a mind reader. The most successful method for organizing bins is to start with a bin for each field tape and then duplicate clips and move them to content-based bins afterwards. This allows you to print a copy of the contents of each tape as it is logged and keep a central notebook of all the tapes. You can go back to the tape months later and know exactly what is on that tape when you archive it or, if the worst should happen and you cannot continue the job on the MC/AX, you will have something tangible to take with you to another edit suite.

Of course all of these strategies assume that you have time to prepare! There will be times when the client just walks into the production house with a box of tapes and literally drops them on you. Now you must log, digitize, and view all at the same time. Some people, not wanting to miss anything, will start to digitize by figuring out how much space is available and how much material is there and digitize every tape from beginning to end. There are two nice utilities called StorageCalc (in your Utilities folder) and Storage Tool by Tony Black (aeblack@randomvideo.com) that can help you with the math. Storage Tool looks at the available storage space and does the calculations based on AVR. This is a good strategy if you are not going to be the editor because it leaves nothing out.

There are times when you must make creative decisions on the fly before you have seen all the material and here you can only operate on your gut instinct and experience. Scan through the tapes as fast as you can and log what appears to be the best and second-best take of every set up. Go back and batch digitize after you have scanned through everything. The benefit to this method is that by the time you get around to actually digitizing, you will have seen everything and can weed out the redundant and superfluous. Be sure to grab some shots that you think are cool and you have no idea what to do with them — they may be the most useful for tying sections together with graphics.

By far the worst outcome of this scenario is when the client tells you exactly what they want, exactly where it is and that is all you digitize. Don't be fooled! You will be looking for shots on tape all day and you will lose almost all of the power of your random access tool. It is imperative that you see everything. They brought the job to you because they trust your judgment, but if you can't see everything then you can't make the creative leap they expect from you under pressure.

## Batch Digitizing

When you finally get to the digitizing step you need to have a standard procedure and follow it closely. You are responsible for technically reinterpreting this footage for everyone else who is going to work on this project. Make sure you can justify every level you adjust. The last thing you want to hear while you are editing furiously is "Does it really look like that on the tape?" There is a button on the Media Composer just for those times, called Find Frame, which asks for the original tape to be mounted and then fast forwards to the exact frame. If you have a nervous roomful of people or a client who is inexperienced with compressed images you may want to map this button and keep your tapes close by. This is an inevitable slowdown, but if your client is unsure, you should take the extra effort to reassure them, especially at the beginning of the project. As they learn to trust you and your judgment, they should request this less (unless there really is something wrong with the originals).

Also keep in mind that the Avid allows you to save those video levels on a tape-by-tape basis. Those video levels may be the ones used when it comes time to finish. Chapter 8 discusses the setting of levels.

A digital project is never really finished. The mess you leave today will come back to you in a month or two as the clients come back for version 2! Or the next editor will curse you loudly as they try to figure why you put things in random bins and left all of those unnamed clips and sequences. Mastering the first part of this workflow will impact the rest of the job and make the difference between always rushing to catch up as you search frantically or staying one step ahead, concentrating on creativity when it is most required, the editing.

## EDITING FLOW

This is the part you should give thanks for everyday, freeing you from the restrictions and limitations you have been struggling against since the first time someone showed us how to load a tape or pull apart a splice. Editing is also the most intimidating step because, without the restrictions, many of the excuses for not being perfect have long disappeared. Some editors fight back by quickly claiming something just can't be done, but be careful, the next editor that works with this director may know how to do it! This is the double edge of the new tool: You are always struggling between what you know how to do fast and confidently and that nagging feeling in the back of your head that if you just had a half an hour you could figure out a better way.

Part of this anxiety comes because we are still trying to use the new tools in the old way. Sometimes we focus on the one best way to do something and bring all of our old skills to bear on the new technology. These skills were honed and perfected because of the demands and limitations of the old tools. This is one of the biggest reasons why experienced editors don't want to spend the time to learn all the tips and tricks. That seems too technical and they just want to edit.

It is easy to get bogged down in option keys and which direction to drag the mouse and it is the last thing you should be concentrating on at the beginning. What you should be thinking about is how to take what is truly valuable and brilliant about the way you work now and translate it to the new way of working. In the process, you should take a long, hard look at those skills you have developed only because they were necessary and discard them if they are no longer so. I used to be proud of how fast I could thread a one-inch machine. The one good part about threading a one-inch machine is that it gets you out of the chair, so you can stretch and think for a minute. That is the real importance of the skill and the valuable part to keep!

There is a lot of personal choice involved in discovering the "right" ways to do particular functions on the MC/AX. The more you work on the system, the quicker you can identify when a technique is not as good as another in a specific situation. Remember the technique or feature that is easiest and most intuitive for you right now. Get one way down and feel confident about that. The rest will come later, but don't let your fear of not knowing the absolute best way paralyze you and keep you from using the way you know. I used to pretend to feel ill, then grab the manual and read it in the bathroom.

The more complicated techniques are valuable since, at some time in the future, you can count on getting a job that will require them. At the beginning, focus on the content of what you are working on and don't let the machine get in the way. The faster you can reach this stage of transparency with the machine, the better you will feel about the new tools. Only then should you think about memorizing the more advanced techniques.

## Refine, Refine, Refine

Do what you do best, edit, and let the machine do what it does best, give you access to your images and allow you to see what is in your head faster. For some, if not most editors, the vision they have comes about through refinement. They have a rough idea of where the scene is to go or the best flow for the material and they start off that way only to find something better, or something missing. Unless you are working with an incredibly tightly scripted show and very little footage, there is a lot of exploration and discovery as you get more familiar with the material. Just like the speed of changing a one-inch reel becomes less important, the requirement to get a cut absolutely perfect on the first try also becomes less important. The digital nonlinear technology is perfect for this type of approach.

The first pass on a sequence is truly a rough cut. Although we have used this term in the past, it usually meant that you tried your best to get everything perfect and afterward there were some changes. Now it means more precisely that you have roughly assembled the parts and are planning a series of creative refinements

I am tempted to not show my first pass to anyone. In particular, I don't want to show it to producers who are not familiar with the possibilities of nonlinear. They would think I was the worst editor they had ever worked with and think about firing me on the spot! I watched a young producer squirm as their VP of communication saw their first rough cut. It had no effects or graphics and had burned-in timecode. They explained very carefully about what they were about to show and the VP still blew his top. It took everyone in the room to reassure the VP that this was all perfectly normal.

Your first nonlinear assembly should be even rougher than this. Many times just taking all the good takes and lining them up one after the other should be your first step. I have worked with clients who were always vaguely unsure if they were using the right take, even after going through an exhaustive review. This is a legitimate concern, especially since every shot depends on all the shots around it. Changing shot seven may make you reconsider shot fifteen. In the past, there was a lot of tape or film shuttling and this was a good way to remind yourself (and everyone watching) what you had to work with. It was also a good excuse to occasionally just look at all the takes one more time. Again, this is one of the valuable old skills that should be brought along to the new tools. Do not discard the time spent watching and thinking because you are not forced to view the material over and over.

Take the circled takes and string them together. Take all the best cutaways and make a true B roll that you can use as a source. Use the match frame command, not to extend your shot as so many beginners do, but to remind yourself what came after your original cut point. Use the time gained by not having to dub or rip apart splices to think, watch, and explore. You may not end up cutting faster, but you will cut better and be happier with the end result. Refine, refine, and refine. Start with big chunks and whittle them down. During the first several passes work for content, then work on flow and timing (if you can separate them).

Use a strategy that incorporates each step only when it is necessary. You should not be worrying about the graphics when you are still trying to work out the soundbites. Some people call this stage the radio cut. Finally, go for the effects and the smoothing of the audio mix. This particular order is totally up to you. If you cut with a full sound mix as your style, then put that as the second pass, but don't try to do everything at once.

You will find yourself watching the scene or segment over and over. In the past you may not have taken the time to do that until you were almost done. Look at the big picture, pull yourself out of the details until the final stages where they belong.

There is room for reconsideration and the type of visual juxtaposition that you might never have thought of unless it was right there in front of you. The ability to hold a scene in your head is a very valuable skill, somewhat like memorizing the opening 50 moves of a chess game. But that is a mental skill not a visual one. The instruction I give beginners as I see them staring at the screen is, "Don't guess!" Don't create scenarios in your head when you can create them with your eyes. The speed with which you can put two disparate and unexpected images together is another tool for shaking up the other side of your brain. Work beyond the everyday logic and justifications you would have used before even attempting a sequence. Seeing new things with your eyes and not your head will inspire you.

## WORKFLOW OUTSIDE THE SUITE

A really good producer can foresee what parts of a project should be left to an editor at an NLE workstation and which other experts should take care of. There are two basic reasons to send parts of the project out of the suite. They are time and expertise.

## WORKING WITH GRAPHICS

If the graphics can be done in tandem with the editing then, when they are needed at the end of the finishing stage, they can just be inserted. Contrast this with the producer who gets to some midpoint and says, "Lets come up with 'a look' now." All editing stops and you become a graphic designer, a 3-D animator, and an expert on four or five third-party programs. Some editors like this new challenge and the creative control and more power to them. The editor is probably the best person to understand how all of the effects and graphics should be incorporated into the show and understand the right formats and work parts necessary to make it happen. But the editor must also make the correct estimate as to how long the third-party design and render time is going to take on any particular computer they are working with. A wrong guess or the failure to build the time into the overall schedule will grind everything to a halt while they watch a revolving cursor and decide whether to come back tomorrow.

Sending all the graphics out takes care of the time and expertise problems, but replaces them with new ones. What if the graphics are not right? What if they are the wrong color or the animations are not long enough to cover the new voice over? The time involved to send them back to the (unsupervised) graphic artist will surely halt the post on your project. A smart producer needs to be able to monitor the bits and pieces being done out of sight just as much as keeping an eye on the progress of the edit. This planning process must also schedule the time to add the graphics at a stage when they will mesh nicely with the editor's final finishing schedule.

All of these concerns are not new. The problem is that in the past, sending the non-editorial bits out to another artist was the only way to get a highly professional finished product. With the onset of the "everything in a box" concept, producers seem to be ignoring their time-honored knowledge and all those who are getting in for the first time have no experience with anything except this new desktop method.

## SOUND MIXING AND COLOR CORRECTION

Many times people ask about color correcting and sound mixing. Yes, the "box" can do it. Who is running the box? Do you play musical chairs as the sound mixer sits down in your chair and fiddles with the sweetening for a while before they let you back in the room? Not likely. Either you are expected to be a sound engineer and a colorist or the process needs to be segmented into parts.

There are some basic challenges with the workflow of sound mixing. You are dealing with a chicken-and-egg scenario. Which comes first: the sound or the pictures? The answer is, "Yes." Thankfully, there is Avid's non-proprietary exchange format OMFI (Open Media Format Interchange). OMFI allows you to send compositions and digitized sound media to digital audio workstations like AudioVision or Pro Tools. This means that there can be parallel processing of sound while the cutting continues. I have held a telephone up to a speaker so someone in New York City could record and use it for a scratch track, but this wasn't something anyone actually planned for. The producer decided that a New York City talent was the only thing that would do and we waited, not so patiently, for the next shuttle to Boston to bring us the master mix. It did help get the governor elected however—no, not Dukakis!

The give and take of the sound track can not be underestimated. That extra few piano notes before a transition might change the length of the dissolve or the amount of time in black. You are shortchanging yourself if you just send the project out for a mix and never see it again until after it is finished and dubbed. There should be time built in to the schedule to tweak the cut with the new music, but this is a perfect world scenario. If you can get your timings and transitions as close as possible using scratch tracks and click tracks, you are way ahead of the game. But wouldn't it be great if you could use the time that you are rendering or waiting for final client approval to coincide with the sound being mixed? What a perfect excuse to wait until tomorrow and get a chance for another "final" pass at the finished product with the finished sound track in place.

The alternative to sending the sound out is, of course, DIY, do it yourself. Lately, there have been some great additions to the sound mixing capabilities of the Avid editing systems. Real-time level adjustment with keyframes, equalization, waveforms, and audio punch-in are all features you have come to expect. The newest release features "Ride and Remember," which allows you to adjust the faders of an external mixer. MC/AX remembers those levels and allows you to further edit them. You end up with lots of keyframes that can then be filtered down to the minimal amount you need. Another new feature is the ability to use Audio Suite plug-ins for all the special effects that you couldn't do before with just EQ-like pitch changing and reversing the sound.

Many people feel that sound mixing will never be complete without sub-frame editing, the ability to make changes at the sampling level, which gives much more control than what is available at 30, 25, or 24 frames per second. Fortunately, Avid digitized sound media is either in the native SDII format (Sound Designer) or AIFC (an OMF encapsulated version of Audio Interchange File Format) and this is compatible with other sound mixing programs like Pro Tools or AudioVision. You can export sequence information as OMFI, open existing audio media, and start mixing. This book will discuss audio importing and exporting issues in Chapter 6.

**Figure 2.3** Version 7.0 Color Connection

## Color Correction

Color correction is by nature a linear process—to adjust shot two, you need to see shots one and three. In a nonlinear project, shots are digitized in an order that reduces tape shuttling and tape changes, but does nothing for allowing the editor to see two juxtaposed shots one after the other until later. There are built-in color correction features that allow a wide range of adjustment in the Avid. Those tools have just been improved with the latest release and include split screen, safe white and black clips, and gamma adjustments. The color correction effect features red, green, and blue adjustments—more adjustment than you have with just a straight time base corrector (TBC). You are probably not a full-time colorist so again there are two questions: Who is running the box and when?

Lets take a look at a few different scenarios:

### Scenario One

You digitize everything for offline and now go back to digitize at high resolution. You set everything to color bars, but there are a few shots that just do not match. You take the time and your best shot at making these simple changes using the color effect and the client signs off and goes home.

### Scenario Two

After digitizing at high resolution, you realize that there are significant sections that require color correction, either because of mistakes in the shooting or just a difference between cameras. Now you have a more serious choice. Is it just the juxtaposition of certain shots? See Scenario One. More likely, entire tapes need to be tweaked. In this case you could create a color effect template and apply it every time that tape is used or use an external color corrector. There are some inexpensive digital color correctors that can be bought or rented just for this purpose. You need to digitize each shot and evaluate it or, after some experimentation, you can find a setting for each tape that is a best light, or tape-based setting. This calls for a specialized skill that is only developed over time.

### Scenario Three

You know you are going to be using lots of different cameras under many different locations and lighting set ups and you plan for tape-to-tape color correction at a professional facility after the final show is output. If you are planning on this method, you should also plan to master on a digital tape format so there will be no further loss of quality during the tape-to-tape transfer. If you have it in the budget, this is the way to go. It solves the most difficult color problems with the talent of an expert that does not affect the editing schedule until after the project has left your hands.

## RENDERING

What would a digital nonlinear project be without rendering? A lot faster. Part of the strategy of this medium is knowing what to render and when. Are there times when you don't need to see all the layers? Take advantage of them and move your video track monitor to view just the minimum. Sure, you need to see the text on V3, but not the color gradation on V2, so render the dissolves on V1 and leave the other tracks unrendered. The dissolves play fine and the titles come up in all the right places and are spelled right. When possible, you can command-click on a video track monitor and turn on video solo to view just one effect, which will now playback in real time. Of course you need to explain what you are doing and use your best judgment. Determine how much can your client see (or not see) before they start to get cranky and you have to stop and render everything to show them what it will really look like.

Think about the kinds of changes that force you to rerender and whether you can make a series of trims and keep everything intact that is already rendered. These trimming considerations are more advanced skills and require that you ensure that all the affected tracks are active when you make changes with lots of rendered effects. This is why you want to get as far into the project as possible before you start to add the effects. You have enough to keep in mind without worrying about the rendered layers at the end of the show when you are still refining the opening montage.

Just how much do you really need to render? Most people render too much, way too much, and it is all because of misinformation or ignorance. Think about the minimum necessary to render and the minimum to monitor. The less you see, the less you wait. Will the machine render on the fly and show you the effect when you are only concerned about content? This book will discuss rendering in more detail in Chapter 10.

## FINISHING

Throughout this process there needs to be one thing in the back of your mind: How are you going to finish this thing? Are you going to a linear edit suite, the negative cutter or does it stay "in the box"? Good form means that no matter what the decision for finishing will ultimately be, and the producer will at some point be forced to tell you, it shouldn't make too much difference. This means you need to name the tapes for an EDL, but make the names obvious so that you or your assistant can find them again easily for redigitizing.

The amount of time you spend on effects should be more concentrated on the show that you know will be redigitized on the Avid for finishing. Otherwise, you are just making a visual guide for re-creation by the online editor. This is a decision about the amount of refinement and a focus of time and effort. It is not unusual to spend an inordinate amount of time on the final 5% of a project if you are finishing. This time would normally have been turned over to an online editor and that would have been their sole job. It is this last stage that most beginners completely underestimate.

There are several ways to save time and money at the final output stage. If you can keep the project in-house, then you can reap extra profits. You can charge for the extra hours involved in finishing instead of figuring in the $400 per hour cost of turning the project over to someone else. Investing in the top AVRs and fastest drives can pay off for keeping everything in the Avid. But to truly save money in the finishing process requires a sign-off from your client that there will be no further changes (get it in writing!). If the project must go to a linear online and as long as the client realizes that you can save hours by manipulating the EDL, then sign-off is the key to the true "auto-assemble." If the show is mostly straight cuts and requires many source tapes, this time savings will be significant.

## KEEP THE AUDIO

There is no compression of your audio when you digitize, no matter what your video looks like. You can use this audio in the linear edit suite—you just need to get it there. The most common method is to record the Avid audio to a timecode-controlled DAT, put it to digital videotape or audio tracks three and four of a Betacam SP. You can lay the audio to the new master tape in the linear suite and then make your EDL video only. You can keep the mix you create in offline and just keep an eye on video levels and effects in the linear suite. Chapter 7 describes this in more detail.

The last important step when you are going to finish on the MC/AX is the tape itself. You need to black a tape and, unless you have two decks, you need to black the tapes for protection masters as well. You don't really need to black a tape if you know how to use the assemble edit function available on the later versions of the MC/AX, but it simplifies things marvelously. Chapter 11 discusses blacking.

## PLANNING FOR CHANGES

When everything is finished, the product is shipped and the check is in the mail, you must concern yourself with one final nightmare scenario. What if they come back? Well, if they are bringing new work this is a good thing, but if they are coming for changes you need to be ready. Can you charge for the changes? Whose fault is the typo? (A good reason to get them to type all the text!) But the real trick is being able to make the client's changes quickly and painlessly and they will certainly keep that in mind when they are projecting costs for their next project.

How are you going to archive this project? Don't be tricked into thinking that an EDL will do the job. You must back up the project file and all the bins. Even if the software changes, you should usually be able to bring any job forward even if it is a year or so later. Large projects require large backups and it may be time for you to invest in removable storage like one of the new, fast, removable formats like SyQuest, or an Iomega, or even a recordable CD-ROM just for the bins and graphics.

Consider also whether you need to back up the digitized media. If you are finishing on the MC/AX, you should build into the cost of the project the amount of time to back up overnight and the cost of an archive DLT. The DLT technology gets faster all the time and is the only way to realistically back up such huge amounts of data. If you have spent the time to color correct and render and everything is just right, then you should spend a little extra time and save it. If you have a day's warning you can have the entire project ready to pick right up again by restoring it the night before. The potential downside is that, when the technology changes, it will change video quality faster that anything else. Eventually, video-digitizing boards will change and your new video will most probably be incompatible with the old video if you have upgraded. Hopefully, there will a conversion utility, but the quality will never be as good as if you redigitize. Not everything ages as well as wine and cheese.

So plan for an adequate amount of time to review and make changes as there will be more of each in a nonlinear project than in a linear project. Have a plan to back up and restore media and twiddle your thumbs occasionally. Everything takes longer than you think it will, but expect the slowdowns and crash edit times to happen in new places.

Workflow encompasses much more than just input, edit, and output. It is full of variations and possibilities and is guaranteed to be unpredictable. Plan for it and be ready for the places where the changes could cost you money and time. If you understand the process, you will be better prepared to avoid the traps and exploit the advantages of the new technology.

# 3

# *Intermediate Techniques*

It may come as a surprise that most beginners at Avid editing make the same basic mistakes. I don't mean mistakes caused by the software being too difficult, but mistakes from working hard to grasp some fundamental ideas. They are sometimes crucial mistakes, like not knowing exactly what to back up and then trying to restore a project with no bins. Sometimes it is a subtler mistake, like not using the power of a new tool because "That's not the way I work." You may be missing a huge opportunity to improve your speed and understanding.

I have heard people say that the Avid is difficult to learn and that the interface has a steep learning curve. This is only partially correct. You can be mousing around the screen in only a few hours and really editing by the end of your first day. But as with any professional tool, you want it to go faster and do more and the Avid interface rewards this. The more times you resolve different post-production needs, the more you can use the functions you already know in new and different ways. There are very few people who use everything the software has to offer. If you have a particular task to perform, there are the tools designed to facilitate that task in a straightforward way. If you need something a little different, there is a lot of room for variations. It is the variations that take the time to learn and are the most rewarding.

The most basic mistakes are made right at the beginning, when editors are still trying to learn how to navigate through their material. There is a lot of translation going on between where they want to go, how they used to do it, and the two or three techniques they know how to use. They end up settling for the dog paddle before they have mastered the breaststroke, the crawl, the backstroke, and the sidestroke. They will always poke along unless they unlearn the method that wastes energy and, lets face it, gets you there without much style.

## USING THE KEYBOARD

What you may have guessed by now is that I am referring to the overuse and abuse of the mouse or trackball. This is where you should start to improve your technique. First, use the most obvious way, the mouse. This helps you get over the beginner's problem of trying to remember what you want to do next and where it is on the screen. After this beginner's stage, you instantly forget how magical it all is and want to go as fast as possible. Then you must cast down the mouse! Use your keyboard!

The best favor you can do for yourself is to force yourself to use the edit keys and keyboard equivalents as soon as possible. If you haven't put the colored keycaps or stickers on your keyboard yet, you are missing a whole world of speed. Look at the command key equivalent for the functions you use the most and think up funny little ways to make them stick in your head. Command-Z to undo and Command-S to save should be comfortable before your first day is over. Then start to use Command-W to close windows and Command-A to select all. Use Command-B to go to Capture Mode and, as an on-going project, memorize the Tools menu. Mark in and out should be mostly done from the keyboard so you can keep your material rolling and mark on the fly. Trimming can be done several ways from the keyboard.

You don't want to give yourself a repetitive stress injury, though, so always try to make the environment as friendly as possible for your wrists. Get the keyboard at the right height, get a wrist pad if you need it, and by all means, give your wrists the rest time and exercise they need to keep functioning. Use the extra time gained with these keyboard techniques to watch the sequence back one more time and think.

## Custom Keyboard

In Media Composer there are many ways to personalize the keyboard, but they should all be created organically from observing the functions and keys you use the most. You can take any button from the Command Palette and put it on any key. I recommend learning the colored keycaps and only modifying the function keys or the shifted functions of keys that make alphabetic sense to you.

**Figure 3.1** A Customized Keyboard

**Figure 3.2** The Command Palette

When you are looking at the Keyboard Setting, hold down the Shift key. You have a whole new keyboard to create! You can put the Render button on Shift-R or the Subclip button on the Shift-S. There are some functions on the pull-down menus that do not have the keyboard equivalents or buttons. These functions must be mapped through a special combination of keystrokes in older versions of the software. While looking at the open Keyboard Setting, hold the Option and Command Keys down at the same time; you will get a cursor that looks like a small pull-down menu. Click on the function key in the Keyboard setting where you want the function to go. Then go up to the pull-down menu and choose it. The initials representing that function will land on the function key. This whole procedure becomes a single check box in version 7.0 in the new Command Palette. You may have made the same mistake I did and didn't learn to touch type in high school; who knew about e-mail then? Once you feel comfortable with the screen real estate, push yourself to resist the mouse and keep your hands on the keyboard!

If you have a custom keyboard, be aware that it works best for the version of the software you were using when you created it. If you go to an earlier version of the Media Composer, your keyboard, like all User Settings, may have features included that do not exist in the earlier version. You can usually go forward with keyboard settings but this has been known to create odd, unrepeatable problems. It is best to take a screen shot of your keyboard setting (use a shareware screen capture program to outline the part of the screen you want as a PICT file), print it out, and make the custom keyboard again. If you are moving from a Media Composer to an Avid Xpress your custom keyboard may work, but again, there will be features that just don't apply.

## J-K-L

Use keyboard control for shuttling. Some love the MUI or the AvidDroid because it feels like what they are used to. Those boxes are better than using the mouse to shuttle or drag through the timeline to view material quickly. Programming the extra function buttons can give an edge to the film-based editor with a yearning for the flatbed systems. Some like it especially for shuttling tapes during the initial viewing stages of the project. But if you don't have access to that extra hardware, the keyboard can be just as fast after the material has been digitized.

Getting through your vast amount of material quickly and then fine-tuning the details can all be done with the J-K-L keys. After you get adept at using the jog buttons to fine-tune, then stop using them! Hit the L key and move forward at 30 frames per second (fps) then hit it again and speed up to 60 fps and then again on up to 90, 150, and 240 fps. (PAL numbers will be multiples of 25 and film is 24.) Eventually the sound will cut out (thankfully) above 3 times normal speed. Hit K or the space bar to pause. Now use J to go backwards with all these different speeds.

Soon you will find yourself cooking through the material at double or triple speed and following the script. Surprisingly, you can understand what they're saying and can work consistently at the higher speed to fly faster through the material without the distracting rollbars on the screen you would see with tape.

This is a linear way of moving through the material. J-K-L is an example of how the computer has improved upon an old technique by just making it easier and faster. Ideally you type in your timecode logs and go to the correct place instantly — that is taking advantage of the random access powers. You search back and forth once you get there. But if you really need to listen and watch the material, there is no replacement for just shuttling through, especially if you were not the person who logged it.

The true beauty of J-K-L comes when you want to make that final editing decision. Holding down the K and the L at the same time will "scrub," move slowly, with audio at 8 frames per second (5 fps PAL). You also get the familiar analog sound effect where the pitch drops and it sounds like Satan has possessed your talking head. Most people like this sound, especially if they are used to editing audio in an analog environment.

## J-K-L for Trimming

So using the keyboard, you can move incredibly fast and slow without taking your fingers off the keyboard. Just wait until you get to the left hand. But first let's take J-K-L to its most powerful (and potentially dangerous) use — trimming. The first time you try this there is a little surprise — while watching the show play back, you are trimming! There are two important benefits to this. Don't scrub and then mark the points for your edit. Make the edit first and then scrub the trim. You will be able to make the fine decisions with the shot already in place. And second, if you mistakenly make a clip way too short, like say by a paragraph, you can lengthen it and listen to the audio at 90 fps at the same time. This is an excellent way to hop through the sequence, one transition at a time, and tighten everything just a bit before moving on to the next level of refinement, while always keeping the hands on the keyboard.

### Scrubbing and Caps Lock

You can achieve a choppy, digital scrub sound by holding down the Shift key and using the jog or trim keys. If you like this better than J-K-L, it is precise and still has its place. But this technique should be a last resort at the very last stage of determining a trim, not the first choice. It is good to know first because you can always fall back on it, but you will soon see that as you speed up with the rest of the interface that, as a main technique, this becomes a dog paddle.

There are some other restrictions with using the Caps Lock key. First and most important, the Caps Lock key takes up a lot of RAM; a lot more than you would think. Maxing out the RAM may cause unusual errors or keep you from seeing certain real-time effects when they are definitely there, like many dissolves in a row at AVR 75. Now that systems are shipping with triple the RAM of the original configurations, this isn't as much of a problem. But it is something to consider if you have an older system and less RAM. Also, RAM is much cheaper than it used to be. If you can, put this book down, go to the phone, and order another 64 megabytes or so. This is the cheapest upgrade possible.

The second problem is that leaving the Caps Lock key down is incredibly annoying. OK, you've escaped from having to hear high-pitched voices screaming backwards when you abandon tape, but now you have blip, blip, blip all day every time you change position in the time line. Stop that! Turn it off!

There, did I sound enough like your mother to get your attention? Hold down the Shift key when you need it; when you don't, it automatically turns itself off when you remove your finger. And start using J-K-L.

## AUDIO MONITORING

Many beginners don't see the connection between what they are viewing and why they can't hear the sound anymore. Watch the small speaker icon next to the audio tracks in the timeline. The solid, completely black icon means that you are monitoring that track. Being able to turn audio track monitors on and off selectively means you can concentrate on just the sound effects or make five versions in separate languages and monitor one language at a time. But it also means you always need to keep an eye on which tracks are being monitored because they could be different from the tracks you are editing.

The risk of using audio scrub is that you can only listen to one track at a time while you are moving slower or faster than 100% speed or moving backward. This is why you can be moving forward at 30 fps listening to music and voice and then when you reverse, you are left with just voice. In general, this is what you want when you are doing fine audio cutting at off speeds. Having all 8 audio channels playing at once is distracting when you just want to find the

**Figure 3.3**   Audio Monitoring Icons

"blip." Instead of turning all the monitors off and back on again, it is done automatically when you slow down below 100% normal speed.

Look closely at the monitor speaker icon. One of them will be hollow, represented just by an outline instead of solid black. This is the track that you monitor at the off speeds. You can move this icon to another track by Option-clicking on a solid monitor icon.

## THE OTHER HAND

Now let's cover the other hand. The E and the R keys should be the main resting place to mark in and out on the fly (or I and O if you are reversing all this). Using the keys to mark in and out allows the video to keep rolling while you continue to search and change your mind. Combine this with Make Subclip mapped to a function key and you can fly through material and organize at the same time. Command-4 on Media Composer takes you from the bin back to the source window to subclip and names an entire bin without taking your hands away from the keys. Alternatively, use the thumb for the V or B keys to actually splice or overwrite the material into a sequence. If you are working from long source clips with many different shots in one master clip, like a film transfer, you can literally type your show together. Use the mouse mainly for loading clips into the source monitor (although the Return key will do that too) and accessing the graphic functions of the timeline.

## MULTIPLE METHODS TO SOLVE ONE PROBLEM

There are ways to do the fine adjustments and still get the analog scrub sound, but with the mouse. It's called the mouse jog and it's a finer scrub than J-K-L. This brings me to another fundamental tenet of using the MC/AX. You could use multiple methods, one after the other, in order to fix a problem. Editors who consider themselves novices may have acquired one or two methods that they use under all circumstances. The more experienced editor uses a method of trim to get close as fast as possible, depending on their particular timeline view and level of sequence complexity. The editor then reanalyzes the problem and easily switches to another method to put the final polish on the fine points. The "right" way is the way that accomplishes what you want in that particular situation in the fastest way possible.

## NAVIGATING NONLINEARLY

Another beginner's challenge is not to think linearly when jumping large amounts of time. Beginners make a compromise and move at the speed of human comprehension in order to grasp a particular point in the material or listen to the performance. But what if you just want to get there as fast as possible? I see beginners actually dragging the timeline through their material or sequence to get to the end! It's random access; get random. Jump to the end with the End key and jump to the beginning with the Home key. (These can be mapped in version 7.0.)

## FAST FORWARD AND REWIND

The fast forward and rewind buttons are useful if you just want to jump to the next or the previous edit but these buttons are usually left to the User Setting default of being "track sensitive." This means that the fast forward and rewind buttons jump to the next edit that uses all the tracks that are highlighted in the timeline. If video track 1 is highlighted then you jump, in a sequential way, from cut to cut on video track 1 only. Turn on all the tracks (Command-A with the timeline highlighted). Now you jump to every edit where all the tracks have a cut in the same place. No straight cuts on all your tracks in the same place? With all your tracks highlighted, fast forward jumps all the way to the end of the sequence. That can be confusing since there is no easy way to get back to where you were in the timeline. No undo for jumping to the wrong place! You can reverse this default Composer setting instantly by holding down the Option key in combination with the fast forward or rewind keys. If you find yourself jumping to the end of the sequence a lot by accident, then in Media Composer, you can change the Composer User Setting so that it jumps to every edit regardless of which tracks are highlighted. You can change the settings to jump to every locator too.

## USING THE TIMELINE

Modifying fast forward and rewind still misses the larger point, which is that you don't need to step through lots of edits just to get through the timeline. If you want to get near the end, just click there! But more importantly, you need to be able to see where you are going. Learn to change the scale or view of your timeline quickly and easily.

Version 7.0 of Media Composer and all versions of the Avid Xpress incorporate a drag bar to resize the timeline. This is marvelous and intuitive. Drag the slider to the left and the timeline compresses; drag it to the right and it expands. You have fantastic fine control and can also change the view by large amounts quickly with the same function. But don't neglect the focus button (H), keyboard equivalents (Command-[ , Command-], and Command-/), and Zoom-In (Command-M) either. Most of these functions are on the Media Composer only because one of the ways the Avid Xpress is simpler is by reducing the amount of keyboard equivalents. The Focus button (H) is especially useful for troubleshooting small problems like flashframes. Command-M allows you to drag a box around the area in the timeline where you want to zoom in. This technique is very useful for pinpointing a segment that needs more refinement.

There are so many easy ways to change the scale of the view in the timeline because it is vital to using the power of the system. The timeline is not just a pretty picture. It is an important tool for navigation and should always be sized to fit your needs for that moment. Zoom in to do some fine trimming; zoom back a little and look at the whole section then zoom in somewhere else to fix a problem. You should be considering the scale of the sequence that your timeline is displaying at every stage of work.

## JUMPING PRECISELY

Even though you may see precisely where you are going, you may not always get there the fastest by just clicking the mouse. You need to combine the power of the random access navigation of the timeline with the precision of the fast forward and rewind keys. How can I jump huge distances in a single bound and still end up on the first frame of the cut? There are a series of modifier keys without which life as you know it could not exist. The most important is the Command key. When you click in the timeline and hold down the Command key, the blue position bar always snaps to the first frame of any edit on any track. It also snaps to marked in or out points.

If you hold the Command key down while dragging your cursor through the timeline, it snaps to the head of every video and every audio edit. If you click anywhere in the timeline with the Command key held down, you are guaranteed to land at the head of the frame or a marked in or out point and can eliminate missing a frame here or there and creating flashframes. You must still combine this with the timeline zooming techniques, especially with a very complicated sequence. You may be surprised that you have snapped to the audio edit on track 7 which is three frames off from the video edit on track 1 you really wanted.

Combine the Option key with the Command key and, as the term option implies, you get a variation of snapping to the head. On the Media Composer, Option-Command snaps to the tail, or the last frame.

## THE OTHER MODIFIERS

The Avid Xpress snaps segments to the edit points automatically when you drag them into the timeline. In the Media Composer there are other important modifiers that snap you to reference points as you drag segments through the timeline. Command and the Option-Command work great for lining segments up to the heads or tails of edits in the timeline, but there are also the Option and Control keys.

In the Media Composer, Option gives you fine, frame-accurate dragging control even though you may be zoomed way back in the Timeline view. In the Avid Xpress version 2.0, Option will line the outpoint of a clip up to the outpoint of an edit in the sequence. This is the Media Composer equivalent of Option-Command. For effects in the Media Composer, dragging while using Control is one of my favorite techniques because it allows me to move things only vertically. If I want to rearrange things quickly and add a multi-layered effect, I can drag V1 to V2 and have confidence that it will not move a single frame. Although the Command key has some of this functionality, if there are inpoints or outpoints close by, you may accidentally snap to those too. The Control key makes sure that you go only vertically no matter what.

## SPEEDING UP DRAGGING

The final way I improve dragging segments in the Media Composer is by turning off the four-screen display that appears as soon as you start to drag a segment. The four-screen display actually takes a second or two to draw itself, longer if you have multiple layers displayed. I find that this delay slows down my thinking if I am really flying along. You can turn off the four screens by getting into the segment mode, clicking on a segment, and, while still holding down the mouse button, pressing the Shift key and dragging the segment just a little bit. The Mac beeps and the four-screen display stays off until you reverse the process or restart the software. In version 7.0, this is a checkbox in the Timeline User Setting and much easier to implement.

## THE IMPORTANCE OF THE LAST FRAME

The MC/AX works from the film model that insists that every frame exists and is important to duration calculations. The video model actually allows you to have an outpoint and an inpoint on the same frame of timecode. How can two shots use the same frame on the master tape? Don't ask, in linear video editing they just do, and video people are occasionally confused when the MC/AX counts every frame separately. Mark an in and out on the same frame. What is the duration? With the MC/AX the duration is one frame. If you are looking at a frame and mark it as an inpoint, then say "Go 15 frames from here" by typing "+ :15," when you mark out, you will have a duration of 16 frames. You have said, "Take this first frame and 15 more," which makes perfect sense to a film editor and seems suspiciously like a bug to a video editor.

There are two ways to deal with this fundamental difference. If you fast forward to the beginning of one edit (or Command-click in the timeline) to mark in and then fast forward to the next video edit to mark out, how many frames do you have? You have one too many. You have included the first frame of the next edit. If you do an Option-Command click, then you are at the last frame of the shot you want (or just back one frame). Your inpoints and outpoints are exactly equivalent to the duration of the shot.

There is an easier way to do this, of course, and that is using the Mark Clip button, which is also track sensitive. This technique avoids the entire problem by just taking the first frame and the last frame of the edit you are working with. Leave the next edit alone!

## TRIMMING

Beginners bring linear thinking to the trimming process as well. This is the biggest potential mistake you can make on the MC/AX, short of deleting all your media. The first place I see this is when beginners misuse the match frame button. I would like to rename this the "Fetch Button" because the match frame name is too close to the function linear tape editors have been using since the beginning of computer-controlled timecode editing.

The traditional tape method is to get the edit controller to find the same frame on the source material as where you are parked on the master tape. The source tape cues up, you adjust the video levels to match what is already on the master tape, and then you lay in a little more of the shot, usually a dissolve or another effect. You can do this in the Avid as well. This is logical and simple but misses the point. Every master clip that you add into the sequence is linked to the rest of the digitized material. You don't need to go get it because it is already there. Think of the extra digitized material as always being attached to every edit in the timeline all the time.

The only reason to use the match frame button is to call up a shot to look at the whole thing or to make a motion effect. It should be used for reviewing material, not as an integral part of the trim process. Used the wrong way, matchframe is another dog paddle.

## Trim vs. Extend

Going into the trim mode is the more sophisticated, but potentially more confusing, way to extend a shot. It allows you to do more in a complex situation. I personally think that the trim mode is the heart of the Media Composer/Avid Xpress functionality. Master trim and you have mastered the system and changed the way you think about editing forever. Trim is creative not just corrective.

The Extend button in Media Composer is the simpler, more straightforward way of extending a shot. Extend can also be faster than trimming. Mark where you want the edit to extend to, turn on the tracks you want to extend, and then hit the extend button. It will not knock you out of sync because it is a center trim. It trims both sides of the transition simultaneously. I find it most useful for mechanical trims that go to a direct and easy-to-mark point. Any finessing and I go to the trim mode.

## Mastering Trim

The main reason that trimming is so much better than just extracting the shot and splicing it back in is that you have the immediate feedback of seeing the shot in context. Working with the trim mode, your goal should be to stay in sync as much as possible. Now obviously, there are times when you want to go out of sync, for cheating action or artistic purposes, and I'm not talking about that. I'm referring to the skill of understanding the relationship between what tracks are highlighted and what kind of a trim you are doing. Some people get so flustered the first few times they try trimming with sync sound that they abandon it altogether and invent elaborate workarounds that are easier for them to understand. Lots of energy, not much style. This is one of those skills that film editors (those who have actually touched celluloid) have over video editors. It is pretty hard to knock yourself out of sync with a tape-based project. Usually picture and sound are laid down simultaneously. But film editors must learn that whenever they add something to the picture, a trim or a reaction shot, they must add an exactly corresponding amount of frames to the sound track.

With the extra power of the nonlinear world (and film was the first nonlinear editing system!) there is the responsibility of keeping track of sync. The MC/AX does a pretty good job of telling you if the video and audio you digitized together or auto-synched together (matching sound and vision from separate sources after digitizing) have lost their exact relationship. They are the white

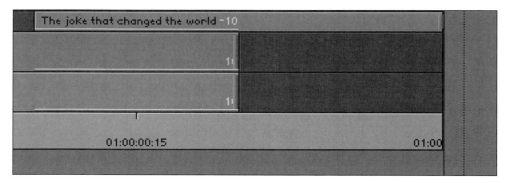

**Figure 3.4** Timeline Sync Breaks

numbers of the sync breaks. I think of them as a silent white alarm that, when I see them ripple across my Timeline, tell me I have most probably made a mistake. The only time I want to see sync breaks is when I have cheated action or I am dropping in room tone.

Before we get much more into the trim mode I want to point out the most important key to use when using the timeline in a true nonlinear way: the Control key. Holding down the Control key allows you to lasso any transition on any track and go into the trim mode at a specific place. In combination with the Shift key, you can make multiple selections easily and, in a graphic way, extend the use of the timeline to get exactly what you want.

## Staying In Sync

The best way I have discovered to think about trimming is to imagine moving earlier in time or later in time to see a different part of the shot. Coincidentally, as you move earlier you may be making a shot longer or shorter. Any trim that adds or subtracts frames, any trim that is on one side or the other of the transition, changes the length of that track and must have a corresponding change on all the other tracks in the sequence. Not all trims change the actual length of a sequence, but the ones that do, the trims on one side of the transition or the other, knock you out of sync if you don't pay attention. This means you must look to the tracks that are highlighted when you decide to add a little video. Don't make the beginner's mistake of thinking that just because you are adding a few more frames to lengthen an action, it is a video-only trim. All the sound tracks must be trimmed if you make the sequence longer or shorter in any way.

## Sync Locks

The easiest way out of this dilemma is to turn on the sync locks. But this is at best a temporary solution. The sync locks allow the MC/AX to resolve certain situations where you tell the system to do two different things. "Make video longer and don't affect the sound tracks." The system adds the equivalent of blank mag, silence, to the sound track. This may be safer than trimming and accidentally adding the director shouting, "Cut!" but it will also leave a hole that must be filled in later. Blank spaces in the sound track are really not allowed! You will find yourself having to return and add room tone or presence so the sound does not drop out completely.

There will be a time when you tell the MC/AX conflicting things. You tell it to make the video shorter, don't change the audio tracks and stay in sync. This is beyond the laws of physics. In this case the MC/AX cannot make the decision for you of where to cut sound so it will give you an error beep and do nothing. Unless the majority of your work is straight assembly with little complex trimming, it is not a good strategy to leave sync locks on all the time. It is very effective, however, when you are locking a sound effect track to a video track. Sync locking multiple video tracks together to keep them from being trimmed separately will help you keep from unrendering an effect.

There is a potential situation where you will be cutting video to a pre-made sound track. The video and audio part of the sequence do not give you sync breaks when you change their relationship. Here, you must be even more careful to be conscious of maintaining sync. Don't fall into the trap of thinking that you can knock yourself out of sync now and later, when you get a chance, go back and fix it. Believe me, by the time you get the chance to go back, you will have created a situation which takes much longer to fix later than if you did it right in the first place.

## Trimming in Two Directions

Trimming in two directions, an asymmetrical trim, is a subtle and very powerful technique. You may have a situation where you must trim a video track longer, but then add extra material to a sound effect or music that would keep you in sync, but ruin the edit. Here, keep in mind that just because you are adding video on V1, you can still trim the black before the clip on A2. Both tracks get longer, but one adds more digitized video, fixing a visual problem, and the other adds more silence, keeping the two tracks in sync.

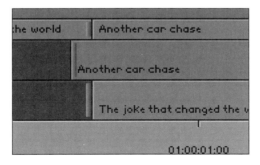

**Figure 3.5** Trimming in Two Directions on the Timeline

**Figure 3.6** Slipping a Shot

## Slipping and Sliding

Beyond those basic situations, most of the difficult sync problems can be fixed by two methods. The first method is to ignore the fact that trimming a shot must make the sequence longer and trim in the center of the transition, basically not affecting the sync, and then use the slip function. Slip and slide are not features in the Avid Xpress. Many people have a difficult time grasping the slip function because it is so tied to the nonlinear, random access concept. In Media Composer you can enter the slip mode by multiple methods. You can double-click with a segment arrow if there is no effect or double-click while in the trim mode as long as the Timeline view allows you to see a black arrow cursor. You can also get there by lassoing the entire clip from right to left. I generally get there while still in the trim mode by double-clicking because I use it in the following method as a second step in a difficult trim situation.

Think of slipping as a shot on a treadmill. The shot slips forward or backward, showing an earlier or later part, but the place in the timeline never changes. A slip will change the content of a shot by revealing new material, but leave the duration of the shot and location in the timeline the same. Because you usually have more video linked to any shot used in the sequence, you can slip that entire shot back and forth.

So if you trim the beginning of the shot ten frames as part of a center trim, you can slip the shot back into position so that it still starts with the same frame. If the first ten frames of shot B are important then slip them back into place. Your center trim moves the frames viewed in the A and B shot of a transition to be ten frames later. Although shot A gets longer and shot B gets shorter, the length of the sequence is not affected and the sync is not disturbed. Shot B has gotten shorter, but after you slip it, still has the same starting frames. I have worked with producers who had edited their programs on the Media Composer for years who had never seen the slip mode! Although it seems complex at first, it is truly a powerful tool used in the right place.

The corollary to the slip mode is the slide mode. It moves the shot neatly through the sequence by trimming the shots on both sides. It affects the location of the shot in the timeline, but not the content or the duration. It is a good alternative to dragging a shot with the segment mode arrows if you are making a smaller change, but not nearly as useful as the slip mode.

## Option-Add Edit

The other main technique to deal with difficult trim problems and maintain sync, especially with many tracks, is the Option-add edit. It is very easy to just lasso all the transitions around the area you want to trim or, with the Timeline window active, hit Command-A (select all), turn on all the tracks, and then go into the trim mode. This can be unpredictable if you have zoomed in on a particular transition to work closely on it and the next edit on V2 is over ten minutes away. You don't really know how you are going to affect this other track, V2, unless you zoom back out to see. You might accidentally add more video to a shot on V2 that you think is already perfect or unrender an effect. The best method for trimming multiple tracks, but with empty spaces on some tracks, is to go into the trim mode and use Option-add edit. This automatically adds an edit in the blank tracks and gives you a place to trim where it would normally be just black filler.

The Option-add edit gives you the handle to stay in sync when trimming multiple tracks. You are left with add edits in the filler—fake transition points, but you can get rid of them by choosing "Remove Match Frame Edits" from the Clip pull-down menu. When you have selected the add edits in the trim mode, you can also hit the Delete key. But my advice is just to leave them. They provide a continuing ability to make changes at that point in the sequence and are also a good indicator that, at some point in the future, you can pinpoint exactly where you knocked yourself out of sync. They won't show up in your edit decision list (EDL) unless you put a dissolve on them—unlikely for add edits in black.

So the most important aspect of trimming is to be aware of when you are going to change the length of the sequence by trimming on one side or the other of a transition and which tracks will be affected. When you grasp these points and overcome the fear of going out of sync, you will have a much more powerful tool and feel much more comfortable with the workflow concept of refine, refine, refine.

**Figure 3.7** Trimming Multiple Tracks with Option-add Edit

## ORGANIZING

One thing that working with a computer forces you to do is organize. If you don't have a plan from the beginning that is easy for you, you just won't do it and will find yourself relying on the frame view to find shots. Although this may make it easier for the client to put their greasy finger on your monitor and shout "That one!," it is slow. The bins that come from the telecine need to be named for scene and take, but after that, everything follows the standard script notation. In a traditional film edit room, organizing has more to do with tracking the physical film and keeping the right scenes ready for cutting. In a nonlinear edit, a film project must keep everything from a scene together. Documentaries or other formats where the form takes shape during the edit must live or die by good organizing of shots in a computer-based edit. I once edited a show with 200 sound bites and no script. In order to use the tools that are available for finding shots, you need to enter the information in a way that allows you to search for it easily.

There are two things to keep in mind when creating bins: tape name and bin size. Generally, the tape name is most important when you are starting to organize since this is what you are initially handed from the field. If the tape names have some criteria for allowing you to trace back shots, for instance, day and location coded into the number used for the tape name, you can more easily follow a series of shots to a common starting point. Also, as mentioned before, you can start with a tape named bin and then organize based on content.

However, if your bins are too large, you are defeating a lot of the benefit of the organizing. It takes longer to open and close large bins, and once they are open, it takes longer to find exactly what you need. Since you will most likely use the Find Bin button frequently, you want it to be opening small bins to speed up the retrieval process. To use the Find Bin function you must have that bin listed in the project window; it needs to have been opened once in the project. Also, you can go directly from the sequence window to the source bin by using Option-Find Bin. In versions 7.0/2.0 you no longer take bins in or out of the Project window; instead you reorganize them into folders. This type of folder-based hierarchy will make the "Find Bin" feature more successful. Again, matchframe is also a useful tool for calling up a shot if you are not interested in the bin. A matchframe calls up the source clip regardless of whether the bin it was digitized in even still exists! All matchframe needs is media on the drives.

## THE MEDIA TOOL FOR EDITING

The Media Tool is a good way to find shots, even if the bin has been deleted, by sorting or sifting and then dragging the shot to a new bin. If the media is online, you can get it through the Media Tool. Most people use it for media management. But on a very large job, or a system with lots of media, opening the Media Tool can be a time-consuming process and some people avoid it completely.

**Figure 3.8** Reusing Standardized Data with Option-click

Early in the process of organizing, if the Media Tool is not too large yet I leave it open to avoid having to keep opening up a bunch of separate bins when looking for a shot. In the Version 7.*x* Macintosh operating system, when there are too many objects in a window, the scroll bars on the sides of the window disappear. This happens to the Media Tool when you have too much material online when you open it. Think of the Media Tool as the largest bin of media you can have at one time and, when it gets really large, only open when you must.

Make use of the customized columns in a bin and try to standardize the way you enter particular information. If you create a column called Shot Type and enter both "Wshot" and "WS," you will have a harder time searching. On the Media Composer, once you have entered a particular notation in a custom column, whenever you go back to that column and Option-click you will get a list of everything you have entered there and can just choose it from this temporary menu. This lessens the chance of spelling errors and speeds up data entry.

So far we have discussed stylistic techniques and the benefit gained is primarily in speed, efficiency, and making you look good. But there are a few common techniques that, if not followed regularly, can create real technical problems. Chapter 9 describes some of these problems, but let's cover a few that can be easily prevented by just following directions.

## BACKING UP

If you are a beginner to the computer you may not realize the seriousness of backing up, but first imagine the cost of losing a day of work. Then imagine losing the entire project. Until you lose your first project, you might not back up on a regular basis. The fact that for about $200 and five extra minutes you could have easily saved all the project information is a very compelling argument. I back up every day, but not to the same disks! Use a separate floppy for every day of the week and you have seven chances to find a useable version.

What should you back up? It is usually difficult to back up all of your digitized media in the middle of a project, but with a DLT and backup software that allows you to back up invisibly while you edit, backing up media isn't out of the question. Backing up digitized media is not really what is important to your job unless you are working with lots of non-timecoded material. What you really need to recreate your job is the project folder on the internal hard drive. Take all of it, not just the bins or just the project icon. Most short-term projects should have a project folder that can fit onto a high-density floppy. If you are working on "History of the World" or need to back up lots of graphics and music tracks from CDs, however, you need some other choices.

The first choice is a piece of software that ships with every MC/AX called Compact Pro. It allows you not only to save your project folder in less space because of the removal of redundant information (Huffman encoding), but Compact Pro also allows you to segment the file and save it across multiple floppy disks. Some people are nervous about this and for a good reason; if one of the floppies fails or grows legs then the entire archive, your project, is unreadable. But I have spoken to engineers who don't trust floppy disks at all, for anything. Besides, this method is tedious and slow if there is really a lot to back up (more than two or three floppies worth or around 3 MB). You can find the size of a folder by clicking on it and then using Command-I to Get Info. If the process is unpleasant, you won't do it.

The best answer is one that has dropped in price recently and that is some form of removable disk backup like a Zip or Jaz drive from Iomega or the equivalent from SyQuest. Now you have 100 MB or more of storage on a single cartridge for approximately US$150 for the machine and US$20 for the cartridge. Attach this to the native SCSI port on your computer, the one that is not accelerated. If you have an external 3D DVE box (the Aladdin), you may have a SCSI conflict until you figure out the correct order and proper termination. The Aladdin 3D DVE needs to be the last in the SCSI chain because it is internally terminated and Zip drives are either 5 or 6 for a SCSI ID. Alternatively, you can send your project over a fast network to a system that contains a drive suitable for backup. On a particularly big job, I backed up to a Zip drive on an Ethernet server on the hour. Even though we lost power almost every afternoon while editing on location, we never lost any data.

Backing up to another hard drive on your system or another computer at your facility may not be good enough. I have had both the Boston Fire Department and Mother Nature ruin two different suites where I was working. (To be fair the fire department was trying to save a historic building.) A particularly successful film assistant I know makes two backups of the project as the film gets close to picture lock. He takes one and his assistant takes one and they take separate routes home. It is Los Angeles after all. Avid Technical Support has a category in their database for recording reasons for equipment failure. Earthquake is one of them. And having a floppy in your pocket also means you are more likely to get paid as a freelancer.

## NON-TIMECODED MATERIAL

In the rush to complete a project, people throw anything and everything into their project just to get it done. Scratch audio is recorded straight to the timeline, VHS material is cued up by hand and CDs are played directly into the Avid without any thought about recreating the job or, worse, starting over again if disaster were to strike. If the time is available, then seriously consider dubbing all non-timecoded material to a timecoded, high-quality format source. The slight loss of quality involved may be the difference between quickly being able to recreate what you have done or matching things by eye and ear. If you cannot timecode all your sources, then seriously consider the DLT for the evening after you have finished digitizing all your media. Remember, a digital nonlinear project is never done, they just run out of time, money, or both.

After saying all this, I will mention a very useful technique for getting CD music into the Avid without actually needing a CD player. If you have a CD-ROM player, you can import the CD track directly into the Avid if the rest of your project is at 44.1 kHz sampling rate. First open up SimpleText or MoviePlayer, the programs that ship with every Macintosh. Within both of these programs, open the desired CD and the track you need and choose Convert. Then go under Options and change to 16 bit, stereo, and 44.1 kHz. All of these must be set right and don't assume that the software will stay at those settings when you shut down for the night. Always check the options the first time you convert audio this way. Then save the file somewhere on a media disk and it will become an AIFF file. You can now import that in the import dialogue of the MC/AX and it will come into the system without ever having been converted to analog. This will save you some tape hiss and preserve the original dynamic range while eliminating the need for a tape deck to get the music or sound effects into the system. However, two weeks from now, when you need to redigitize the project for unexpected changes, you will have to re-edit all these elements back in by hand unless you back up the audio media.

## CONCLUSION

These are a few of the types of techniques beginners don't grasp from the entry-level course. They are usually a combination of changing some of your work methods and taking advantage of some unfamiliar functions. There are many other keys, modifiers, and tips and techniques in the MC/AX, but these are the main areas to concentrate on. Use trim and start to think nonlinearly about structure of the sequence. You will find your speed incrementing in leaps and bounds.

# 4

# Administration of the Avid

Some people are lucky enough to have their Avid system administered by someone else. Hand them this chapter and go back to cutting. However, the day-to-day reality is that most people are responsible, at some level, for their own system. After all, if you have created a difficult situation for yourself, it is you that cannot go home until it is resolved. A smooth running session makes you look good, period.

## ROOM DESIGN

It is easier to administer a system that is set up correctly and put into a room that is well designed for the Avid. There are fewer constraints placed on the design of a digital nonlinear suite than a traditional, tape-based suite, but this doesn't mean you can just plunk the equipment down in a pile! There is flexibility to move all the equipment into another room and keep the bare minimum in the suite or move the noisiest parts like drives. There are technical implications to making cables longer as you will see. Suites need to be designed so they can be serviced and simple maintenance can be done without disruption.

If you plan to use the suite as an online finishing facility, then you should treat it like an online suite. Use an external waveform and vector scope, high-quality speakers, a high-quality third monitor, and seriously consider a patch bay. There is nothing worse at 2 AM than carrying VHS decks around so that you can do two dubs. If connections are made through a professionally installed patch bay there is a better chance the dubs will come out well!

Probably the biggest consideration after editor ergonomics is the noise level. Many people make the mistake of putting all their equipment in one room because the SCSI cables that ship with the system are so short. Longer SCSI cables are easy to purchase at your favorite computer or hobby store, but you should not use them for any reason. These will almost certainly cause weird problems and data errors later on.

The slickest installations I have seen have completely moved the Macintosh and the drives to another room. They have extended the keyboard or ADB (Apple Desktop Bus) cables, and the monitor cables and they have attached or custom installed an external floppy drive. Truly, there are no devices that have fans like the drives or CPU that cause much of the noise. Of course, if you have no central machine room and the editor must load tapes and digitize, there should be a connection for a video deck's video, audio, and control cables. If all the hardware is in another room, there should be telephone links (preferably with speakerphones) between the two rooms for troubleshooting purposes.

The best facilities have gone back to the central machine room design after the initial years of isolated Avid suites. It is important now to have a range of decks available during the course of an edit so multiple decks need to be available. Tying up a deck all day when it is not really needed is just as bad. A central machine room also makes it easier to get to the equipment for support or upgrading. There is nothing worse than working under a table with a flashlight in your mouth, the phone in one hand and a screwdriver in the other. Unfortunately, this happens way too often because of thoughtless room design.

If moving all the parts with fans is too elaborate, I have seen rooms where they needed drive towers and were able to use the longer SCSI cable that comes with a differential ATTO card. You need the differential ATTO card and differential cables or something like Fibrechannel to put the drives any farther away than just a foot or so (the length of the cables that come with the system). The use of differential SCSI led to one support call where the freelancer insisted that they didn't need drives because "they just plugged into the wall." Show your freelancers the layout before they start the night shift!

Many who build rooms with extended ADB and monitors use the Gefen Systems ADB Extender. This device amplifies the ADB signal so that it can be sent as far as several hundred feet. Monitor cables for video and audio can be extended quite a distance using standard audio and video cables but must eventually be boosted with distribution amplifiers.

One of the nicest things about many of the Avid suites I have worked in is the addition of windows. There is nothing better to clear your head than to just stick it out a window for few minutes and having natural light is a nice change. But I must fall back on the admonition that if you are doing color correction or shot matching of any kind, you need to have complete control of the lighting. Color temperature of sunlight changes during the day so that the shots digitized in the morning may not look like those input at noon unless the ambient lighting is indirect and consistent. If you have windows, make sure you can pull the room-darkening shades and keep glare off the monitors

If you are also planning to do final sound mixing, make sure the room has been deadened. Applying sound absorbing foam around the room can do this or just make sure that the room has enough carpeting and wall hangings. Mixing in an empty office is probably a bad idea. Investing in good speakers and a real amplifier pays off quite quickly.

## ELECTRICAL POWER

Personally, I feel no suite is complete without two phones, a trash can, a box of tissues, and a dictionary. You'd be surprised where people set these temporary suites up: attics, basements, bedrooms, storage closets, hotel rooms, boats, bank vaults, and ski lodges. Try to minimize any environmental impact like heat, dust, or jarring motion (like editing in the back of a moving truck). The final and probably most important piece of equipment you need under any conditions is an uninterruptible power source (UPS).

If you are running a system now without a UPS, you should nonchalantly put down this book and run to the phone to order one now. You are living on borrowed time. A UPS regulates your power, giving you more when your electric company browns out or less when you get a spike. If power fails completely, a UPS gives you enough battery time to shut down the system in an orderly fashion and avoid crashing, losing work, and potentially corrupting important media or sequences. A UPS makes your equipment run with fewer problems and you will be able to charge for more productive hours of use.

The real question is not do you have one, but how much of one do you need? They are figured in the confusing scale of volt-amps. The math is not so hard if you can find out what each piece of equipment needs for electrical power for either volts or amps. The numbers are usually in the manuals or on the equipment itself. *The Avid Hardware Installation Guide* lists the power requirements of each piece of equipment. But don't neglect connecting the tape deck. What happens to your camera original if the power goes out when you are rewinding?

Here's how to figure the size of an uninterruptible power source that is sold in volt-amp models:

$$volts \times amps \times power\ factor = volt\text{-}amps \times power\ factor = watts$$

The power factor for computers is between .6 and .7. So you can look at the same equation as:

$$watts \times 1.4 = volt\text{-}amps\ for\ computers$$

*watts* for a Media Composer 8000 with 4 iS 9 drives, a PVW 2800, and a DLT = 1111 watts, so

$$1111 \times 1.4 = 1555.4$$

And because you don't really trust manufacturer specs for the UPS and you buy more than we need to accommodate future expansion, you add another 33% on top:

$$1555.4 \times 1.33 = 2068.7$$

You purchase the 2200-volt-amp model and sleep well at night.

Also keep in mind that before you get a true blackout, you will probably suffer from sags and brownouts, which may cause the UPS to use up some of its battery power to keep you going until the Big One hits. Then, when the power comes back on, you can count on a serious power spike. A spike can cause damage to boards, RAM, and drives and that damage may not show up until days, weeks, or months later as the parts start to fail prematurely. The fact that all power is going through a series of batteries and power conditioners with a UPS before it gets to your delicate equipment should give you a warm feeling in your stomach.

If there is any question about what to put on the UPS, imagine using that device full tilt when the power goes out. Ever see a one-inch machine lose power while rewinding a finished master tape? Not good. A Betacam SP will almost certainly crease the source tape if it is rewinding when the power goes out. Even if all the lights and your monitors go out you can always save and shut down quickly just using the keyboard. In an emergency remember, Command-9, Command-S, Command-Q, and Enter. That sequence saves the project and quits the application to avoid the chance of corrupting the project from being shut down improperly. A UPS has saved me literally a dozen times. After your first serious power hit, what is the real cost of replacing your system one board at a time?

## ERGONOMICS

Human ergonomics has been written about at length in other places so just a quick word about it here. Don't scrimp on chairs. They make the difference between happy editors and editors in pain. Get chairs that can adjust armrests, back and height. Many people swear by armrests and, with a keyboard and wrist rest at about the same level, there is less chance of wrist strain. Keyboards can be put on sliding shelves below the workspace. Keep the back of the hand parallel to the forearm to reduce wrist strain. The relationship of chair, keyboard, and monitor cannot be underestimated. Some editors, like multi-Oscar winning editor Walter Murch, edit on the Avid standing up!

## MEDIA MANAGEMENT

Let's discuss the nuts and bolts, the bits and bytes, of what happens when you put media on your system. The MC/AX is an object-oriented program, which means that many things you do create an object. Digitizing media, rendering, importing, creating sequences and bins all create different kinds of objects. It is the relationship between those objects that allows you to combine things in such interesting ways.

The limitations of desktop computers become apparent when you are dealing with hundreds of thousands of objects. Achieving object counts of this scale is not uncommon and once the MC/AX begins tracking around 150,000 objects, you notice a slowdown. Features do not respond as fast and you find yourself clicking and waiting more often. Buying more RAM can alleviate much of this. Because feature films require lots of footage to be online at the same time, the Film Composer ships with more RAM than the other systems. The more media you have, the more objects you have. If you have several 60 GB media drive towers connected to your system and they are full, you may easily have over 400,000 objects. If you are working on projects that require a lot of effects, you are rendering often and creating even more objects. Cleaning off the unused rendered effects is essential to making your system run at speed. In addition, if you experience slow response time, don't overlook just restarting—it defragments your RAM.

The number of objects can creep upward if you are not paying close attention, or if there are multiple shifts digitizing and they are not communicating. There is an easy way to find out how many objects are on the system now. Go to the Project window, click on the Info button, and click on Memory (or choose it in the first menu at the bottom of the window). There is a listing of the amount of objects the system sees now.

The good news is that there are ways to reduce the number of objects without actually deleting material. Deleting is quick and easy, but if the material has no timecode you are forced to either recreate it later or edit it back in by eye. Being able to clear off drives when they become too full with large amounts of small objects is a good argument for dubbing as much as possible to timecoded sources, but that takes time too. As an administrator, you need to decide about policies regarding the speed and ease of replacing material and when it is appropriate to clean off drives completely. As you will see later, you may end up archiving certain crucial elements and redigititizing the rest.

The MC/AX only sees the media that is in the folder on the media drives named OMFIMediaFiles, 6.*x*MediaFiles, or 5.*x*MediaFiles, depending on the version of software you are using. This folder was created automatically and named by the software when the application was first launched and the drive was used

**Figure 4.1** Number of Objects in System

for digitizing. The OMFIMediaFiles folder must keep this precise spelling and it must stay on the highest level of the file hierarchy. In other words, you cannot put this folder inside another folder and you cannot rename it. If you do, the media inside the folder goes "offline" and is no longer accessible to you when you try to edit. Also, any folder that is put inside the OMFIMediaFiles folder appears to be offline as well.

## THE MEDIA DATABASE

### Pre-Versions 7.0/2.0

In versions previous to 7.0/2.0, the way the MC/AX keeps track of all the media on all your drives is through a small file called the Media Database. The Media Database is inside every MediaFile folder and is continually updated every time something is added or subtracted to the media on the drives. If you change anything in the MediaFile folder, the next time you go back into the software, the MC/AX stops and rebuilds the Media Database for you. This saves a lot of confusion and automates a very tedious task, but it goes to show how important that little file is. If your drives are too full for the Media Database to expand, for instance, you are in serious trouble. The Media Database may not be able to completely rewrite itself because there is not enough space and it will continue to interrupt you until it gets that space. If you are interrupted in the middle of normal editing by the Media Database being rebuilt, then you have a serious storage problem and should instantly clear space from your fullest drives. It is normal to see the rebuilding message however, whenever you have done something on the Finder level to the media files like copying or deleting and now have switched back into the MC/AX software.

Occasionally, as with all computer files that are written to and read from on a regular basis at high speeds, there is the chance of file corruption. This may happen if the drive doesn't update the Media Database absolutely perfectly because of a small drive malfunction. Later, when it comes time to read the

**Figure 4.2**  Media Database

Media Database, there may be problems. The information may not have updated correctly or it may be unreadable.

## REBUILDING THE MEDIA DATABASE

There are two simple procedures you can choose from if you find that you have lost media that you know should be on the drives and you suspect there is something wrong with the way the system is keeping track of your files. You can go to the MediaFile folder and find the Media Database by just clicking somewhere inside the folder and typing "M" (this is a cool Mac trick). You can also use the Find File command to find them all at once. Then, take the Media Database and drag it out of the folder and back in again. This may seem futile, but it is actually telling the MC/AX that something has changed and forces the database to rebuild. If this doesn't work, then you may have to actually throw away the Media Database by dragging it to the trash. Actually trashing the Media Database is a last resort and causes the software to start completely from scratch in making a new Media Database the next time you activate the application. This technique reinforces how important the information is in this little file, since this simple action my cause offline media to become online once again.

The Media Database only knows about the files in the MediaFiles folder and so only counts these files as objects. Anything dragged into another folder or out to the desktop is considered offline. This is an opportunity to manage your media in clever ways because it allows you to keep the media on the drives and still lower the number of objects. There is a wonderful third-party application written by Tony Black called Media Mover (aeblack@randomvideo.com). Avid ships it when you buy a DLT, but you can buy it directly from him over the Web. What makes Media Mover so fantastic is that it lets you quickly and easily organize your media by projects.

Media Mover goes out and searches your media drives and returns a list of the projects it has found. You can then choose to move those media files out of the common MediaFile folder and into a folder with the name of the project. So when a project is done for the night, the night assistant or the post-production manager can take all the media that was generated that day and move it to its own folder where no one should mess with it by accident. Thus, the number of objects is reduced, the system response speeds up, and the drives are simplified for the next project to start in the evening shift.

The next morning rolls around and it is time to start with the day job again. The night assistants, before they leave for the morning, go back to Media Mover. They move the evening project to its own folder and, employing the alternate function of Media Mover, they move back the day project into the MediaFile folder. Simple, neat, and predictable. We'll return to this scenario when we discuss archiving media on DLTs.

### Media Files and Versions 7.0/2.0

The Media Database is replaced in versions 7.0/2.0 of the MC/AX with two other files called msmOMFI.mdb and msmMac.pmr. The MSM type of media file in versions 7.0/2.0 is designed to work on large media servers and handle media in group user situations in different ways. So when media is changed in the OMFIMediaFile folder, the editing software is not automatically updated. Imagine being interrupted every time someone moved something to or from one of your media drives if they were accessing it from another system. In order for the system to look at what is actually in the OMFIMediaFiles folder, you must choose Update Media Database from the File menu or open the Media Tool. After this, all offline media that is really in the folder is linked. This chapter will discuss relinking in more detail later. In versions 7.0/2.0, relinking is more important then ever.

## COMPRESSION, COMPLEXITY, AND STORAGE ESTIMATES

All of this clever manipulation of objects still does not solve the basic problem of running out of space on the media drives. It is only a matter of time before this problem occurs and you should prepare for it in an organized way. There are several ways to tell when or whether you are going to run out of space. The Hardware Tool under the Tools menu gives you a bar graph of how full the drives are relative to each other. This is good for figuring out where to start digitizing the next job, but it does not give you specific numbers to make you feel confident about the whole thing. Prior to versions 7.0/2.0, you could not use the Hardware Tool while in the Digitize Mode because the Digitize Mode marked off all empty space on the target drive that it was digitizing to. According to the MC/AX, this space is reserved just for digitizing and appears to be already filled when using the Hardware Tool during digitizing. In versions 7.0/2.0, the Hardware Tool has been changed and is accurate no matter what mode you are using.

If you need precise numbers, then open the Digitize Tool and choose the tracks you think will be needed the most. If you are working with material that has sync audio, then turn on all the tracks. But if you are working with mostly MOS (silent) film transfer that will be cut to a pre-made soundtrack, then get the estimate with only the video track turned on. Video takes a massive amount of space to store, even compressed, compared to audio files. The Digitize Tool gives you an estimate of how much space you can use. The estimates in the Digitize Tool, however, do not show you how much space is on the drives. This is a common misunderstanding. If you think you have the space to digitize 30 minutes at AVR 75 you might actually try to do that and that would be a mistake. There is, at present, a two gigabyte file size limit on the Macintosh which will change with Rhapsody. You currently can not have any single file larger than two gigabytes on a Macintosh. The moment any file exceeds this size, it drops off the edge of the world never to be seen again. (The MC/AX does not allow you to do this, although some programs will.) Thus the numbers in the Digitize Mode tell you not how much space you have, but the largest possible single file you can digitize and stay under the two-gigabyte file size limit. This means if you have a large file which is digitized at a low AVR it may be too large to redigitize at a higher AVR.

Versions 7.0/2.0 of the MC/AX allows you to digitize one clip across multiple drives and across partitions of the same drives. In Digitize Mode you can tell the system that when a clip is under a particular time length, say 30 minutes, and you need extra space, then jump to another drive and continue without a break. If the clip is longer than 30 minutes, it tells you that you do not have enough space just like it does now. In Digitize Mode you can specify a group of drives which will be the only drives available when the system needs to jump to another partition.

If you select all the clips in a bin and choose Command-I (Get Info), this opens the Console window and gives you a total length of all your clips. This is a powerful way to see whether you have enough space on your drives to redigitize everything in a bin.

Unlike video or film, compressed images are judged by their complexity. A very complex image always takes up the exact same amount of videotape to record. When you digitize an image to disk, it is very important how complex a shot or image is because the more information and detail in frame, the more space it takes to store and the more difficult it is to play back. Playback from a disk-based system is a question of throughput or how much information can be read from the hard drive, pushed through the connecting buses into the CPU and out to the monitors. This is why a lower end system may not be fast enough, because it can not get the information from the drives fast enough to play all the information of a high resolution image in real time with effects and audio.

Some systems actually drop frames, but the MC/AX does not. Instead, you receive the dreaded video or audio underrun error message and playback stops. You need to decide how to get that section to play through and this chapter discusses that later. It is remarkable that the MC/AX will never drop frames during playback. This is crucial to trusting Digital Cuts as a visual reference to compare to EDLs and Cut Lists. Not many systems are always frame accurate for output.

At present, all compression done by the MC/AX uses something called Fixed Q tables. Fixed Q means that a shot is compressed with constant quality. If a shot is very simple, it has few details and not many moving objects. As a result, it is compressed more because the human eye cannot perceive the information is missing. The Motion JPEG compression that Avid uses can easily figure out what the human eye can detect and so does not compress images as much that require lots of detail. In general, M-JPEG compresses color more than detail, again because of the physical abilities of the human eye to see loss of detail before loss of subtle hue shadings. But M-JPEG in its present form in use in the MC/AX is lossy. This means that information is discarded and cannot be recovered. Because of that, there are certain combinations of colors and subtle gradations that do not hold up well.

There are three categories for digitized images: uncompressed, lossless compression, and lossy compression. As you move closer and closer to uncompressed quality, you pay in terms of a large leap in costs of hardware and disk space because you must carry over every pixel of every frame, no matter how redundant that pixel is. Uncompressed images demand faster computers and much more disk space on more expensive and faster drives. Only time will tell what technology eventually becomes the common format, but the ironic situation at present is that although uncompressed is the goal for nonlinear editing, only two of the digital tape formats are indeed recorded component without compression. These formats, D1 and D5, are not for those on a low or even medium budget.

What this all means to you is that if your material is all talking heads, it will be compressed more and take up less space than a show that has lots of slow zooms of engravings, moving leaves, or crowds. This is why all space estimates are given with high and low ranges. If, at some point in the future, Fixed Q falls out of favor, then all the images take up the same amount of space, making storage needs easier to predict, but taking up more storage space overall.

## AUTO-DELETION OF PRECOMPUTES

Predicting available space on media drives must go hand-in-hand with keeping track of rendered effects, or precomputes, and imported graphics and animation that also take up space. Every time you render an effect it takes up space on the drive and even though you may cause that effect to become unrendered or delete it from the sequence, that file still lurks on your media drive. This is actually for a very good reason, but it means you need to pay attention to how full a drive has become even though no one has digitized to it that day.

The MC/AX does not keep track of how many sequences are created since there could be multiple CPUs all accessing the same media or archived sequences that are modified on another system and brought back. There is no way the system could definitively know or make any kind of judgments about the number of sequences created. Therefore, if it was to make very important decisions for you, like deleting unneeded rendered effects, what would happen if you called up a sequence you had spent hours rendering to find that the software had neatly deleted that media automatically thinking you were done with it? Just because you deleted the effects sequence from version 15 doesn't mean you don't want effects on versions 1 through 14. There are too many variables and this is too important a decision to leave up to an automated function at this stage in the technology.

That being said, there is in fact some auto-deletion of precomputes going on under your very nose! But, as it should be with all automatic functions that cause you to lose things, it is very conservative and you may not even notice. The only auto-deletion precompute scenario occurs when you are making creative decisions quickly and removing effects you are unhappy with from the sequence. If you are rendering the effects one at a time and then quickly deleting them, there is at no time any opportunity for the system to save the sequence with those effects in place. There is no record that they will be needed in the future because they have not been saved. Saving happens automatically at regular intervals and when you render effects in a bunch they are saved as the last step before allowing you to continue. Every time you close a bin, you also force a save of the contents. That is why the software only auto-deletes when you are rendering one at a time and quickly removing or unrendering them by changing and tweaking. If a save occurs while an effect is in a sequence, then the effect is not deleted automatically.

Every little bit helps to keep the drives unclogged, but you still must evaluate the amount of precomputes and delete them on a regular basis. This is really not so hard even though it is a little intimidating at first because it involves deleting material that someone (probably the editor) may need if you get it wrong. This chapter will deal with the isolation of precomputes when we look at efficient deleting strategies.

## THE IMPORTANCE OF EMPTY SPACE

Remember the Media Database? Any file that keeps track of all the media is a concern if you fill your drives too full. That file must be allowed to enlarge to deal with the many files you add during the course of editing. If there is just not enough room, you may have media file corruption and eventually drive failure. A good rule of thumb is to leave at least 50 MB of each partition empty. It is also a good idea to erase media drives completely after a job has been completed. Don't get erasing a drive confused with reformatting! That is only a very last resort to save a dying drive.

## BETWEEN SESSIONS

If there are multiple shifts all using the same system then the changeover period may be as simple as shutting down, turning some drives off, turning some others on, and booting back up. Since you want to minimize the actual disconnecting and reconnecting of SCSI cables (they are very fragile and subject to data errors if twisted), this is a good reason to change to MediaDocks. If the media is moving to another system, consider also putting the project files on the removable media so the project can be picked up and continued wherever the drives go. If these removable drives are too expensive, it may be time to begin restoring and archiving projects on DLT.

There will come a time when a project must be moved from one system or facility to another. Also, it may be important to keep only part of a job and free up space for something new. These two situations require a better understanding of the sophisticated media management tools of the MC/AX and a few more steps on a regular basis to keep things running smoothly.

## CONSOLIDATE

The best tool you have for moving media inside the MC/AX is Consolidate. Consolidate can be used for two main purposes: moving media from multiple drives to one drive and eliminating unneeded material. Basically you take whatever you need to move, either an entire bin of master clips or a finished sequence, and consolidate it to another drive.

If you want to move media from multiple drives to a single drive you can consolidate master clips. This will take all of the clips in a bin and copy them in their entirety to another drive. This is fantastic for being able to clear off multiple drives and put everything from one project onto a single drive or onto a series of drives so you can remove them or back them up. This does not shorten material, it just is convenient and takes a complicated media management task and makes it one step. Again, be careful not to overfill any one partition.

If you want to reduce the amount of media on your drives but keep just enough to continue editing the master sequence, then you want to consolidate the

**Figure 4.3** Consolidate

sequence. This will break the sequence into individual master clips and copy just the material required for the sequence to play. This creates shorter versions of long master clips because if the whole clip was not used in the sequence, then the new master clip will now be shorter. You can then selectively delete unused media.

Consolidate is especially important at the end of projects when the final sequence has been completed and it is time to back up. Instead of backing up all the media, you consolidate first and only back up the amount of media that was actually used.

## Consolidating a Sequence

The beauty of Consolidate, though, is the advanced way that it looks at everything that is needed in a sequence and only copies that. There are, after all, other ways to copy media, but it is very difficult to tell at the Finder level which clips are really necessary to play a sequence. Whereas you would have to search all the original drives for the shot you want, the Consolidate function searches them all for you, gathers up only the bits you need, and then copies them to the desired drive.

So you select the sequence and consolidate it and the MC/AX looks at the original master clips to determine exactly what is necessary. If you have an original master clip that is 5 minutes long, but only used 10 seconds of it, then only that 10 seconds will be copied. The new 10-second master clip has the original name and a ".new.01" if it is the first time in the sequence that shot is used. If you use another 5 seconds of the same original master clip, you have a second consolidated clip with ".new.02," and so forth. Consolidate also allows you to specify handles, or a little extra at the beginning and end of the clip, so you can make some little trims or add a short dissolve later. If the handles are too long, you will throw off your space calculations, so be careful. However, if the handles cause two clips to overlap material, the MC/AX combines the two clips into one new clip rather than copy the media twice.

## Consolidating Master Clips

If the idea is to move media to another set of drives, some of which have been used already for digitized media, then be sure to check the option to "Skip media files already on the target disk." You use this if you are consolidating master clips and not sequences. If you are copying an entire bin full of media files to a drive, and some of the media is already there (because you already digitized it there, for example), then you do not want to copy it again. Save yourself some time! There is no need to copy media that is already on the drive and you are not shortening anything when you consolidate master clips. When Consolidate finds the long, original master clip already on the target drive, it leaves it alone, untouched. If the idea is to delete unneeded media, like when consolidating a sequence, then uncheck "Skip media files already on the target disk."

**Figure 4.4** Copying Master Clips with Consolidate

One mistake many people make is that they don't take advantage of a feature that, when working with master clips only, allows them to copy the media, but link it to the original master clip. In other words, you don't end up with a ".new" on everything you will be using from now on. Instead, all the original media is connected to a master clip with a ".old" added and the new, consolidated media is connected to the original master clip without anything appended to the name. You may choose this option if you are planning to delete the old media anyway and so you will not end up, after a few consolidations, with clips that have ".new.new.new" appended.

So if you are consolidating sequences, choose the drives, the handle length, and whether to "Skip media already on the target drive." If you want to delete the original media after you have created the shorter, consolidated clips then check the "Delete media original media files when done" if you are feeling brave. If you are feeling prudent, go back and delete the originals in the Media Tool after consolidation. If you are working with multicam, choose to consolidate all the clips in the group if you want to continue to have all the camera angles available in the consolidated sequence.

If you are consolidating master clips, choose the drives and "Skip media already on the target drive." You probably do not want to delete original media here and will proceed with more sophisticated media management later.

These are the two major reasons to consolidate although as a quick troubleshooting tip, you may choose to consolidate a clip that is not playing back correctly. If the clip plays back better after consolidating to another drive or partition, you may have a drive problem or you may have digitized it to a drive that was too slow to play back that AVR.

### Fixing Digitizing Mistakes with Consolidate

Let's take the example of too many tracks. Perhaps you weren't paying attention when you were digitizing and added video tracks to a voice over master clip. You can take this master clip and subclip the entire length in the source window, making sure that you turn off the tracks that have no use. If you have digitized voice over and accidentally kept the (black) video, then when you subclip the master clip, turn off the video track while it is in the source window. The subclip will be audio only. Highlight that subclip in the bin and choose Consolidate under the Clip menu. The Consolidate function copies only the media you want to another place on the drive. You will have a new subclip and master clip with only the tracks you want and can delete the originals to free up disk space.

You can also use this method to just shorten your master clips after you decide what part of them you really need for the project. In fact, many people like this method as a general strategy and digitize master clips that are quite long, maybe an entire scene or the full length of an already edited master tape. This creates fewer files on your drives for the computer to keep track of, fewer objects for an object-oriented program, and can speed up performance. The only drawback is that you may have a problem if you try to redigitize an entire master clip at a higher resolution on pre-7.0/2.0 versions of the software. The high resolution causes the file size to exceed the 2 GB limit for the Macintosh. Versions 7.0/2.0 allows you to split large master clip media files across multiple drives and thus avoid the single file size problem.

## COPYING ON THE FINDER LEVEL

If your goal is just to move an entire project to another drive, you are probably better off working at the Finder level. If you have been using Media Mover, your job is a snap; just copy the entire folder with the name of your project from each of the affected media drives. It is easy to do a Macintosh Find File (Command-F) and just copy every folder found with the project name. The situation may be complicated by the fact that there may be two different resolutions to keep track of in this project. You may only want to copy the low AVR material and leave the high AVR on the drives or vice versa. Planning ahead helps here and I recommend that you use the Labels function provided by the Macintosh operating system. If you go to the Control Panel for Labels, you can change what the colors represent. Change the blue label to Project X Low Res and green label to Project X High Res. Select all media files immediately after you have digitized them and change their label color. You can sort the media files by label color and copy or delete only the files you want. This also allows you to track down those stray media files that always manage to escape even the most careful herding.

Whether you choose to consolidate or work on the Finder level, you need to keep track of all media. It must be searched for, backed up, copied, or deleted. The number of objects on the system should be periodically checked and unneeded rendered effects or precomputes must be deleted on a routine basis. If your drives fill up, the session stops.

## DECIDING WHAT TO DELETE

A place where an administrator or an assistant must be very careful and yet very efficient is the deletion of unneeded material. Sometimes this can be agreed upon mutually with the editor and you can eliminate all of the material for Show 1 when you are well under way with Show 2. But many times, different projects share the same material and you must be careful not to delete that which is needed by both. Since the Media Tool and Media Mover can both be used efficiently by sorting by project, you must find some other criteria to sort or sift by, protect certain shots, or change the project name of the material you want to keep.

## USING CREATION DATE

Creation date becomes a very important criteria to look for individual shots and is often overlooked by many assistants. It does not work like the modified date on the Macintosh finder, which updates to reflect the last time someone opened and modified the file. The creation date is stamped on the clip when it is logged, so if the shot is logged and digitized on the same day, this becomes a useful heading to eliminate material that was digitized at the beginning of a project.

I especially like to use creation date as a heading in my sequence bin. I duplicate my sequence whenever I am at a major turning point or even if I am going to step away for lunch or dinner. When I duplicate the sequence, it time-stamps it so I know that I am working on the latest version. Then I can take the older sequence and put it away in an archived sequence bin. I have control over the exact times that I have stored a version (instead of leaving it up to the attic) and I can keep the amount of sequences in any open bin to a minimum. Sequences will be the largest files you will work with and, in an effort to have the bins open and close quickly and not use up too much RAM, I try to keep the bins smaller and only have the latest version of any sequence.

If you are trying to determine which sequence is the latest version and the editor is not present, creation date is the best tool. There is always the chance that the editor duplicated the sequence and continued to work on the old one, but you should have an agreement with the editor about how to determine this crucial fact.

## CUSTOM COLUMNS

There are many other criteria you can sort and sift by. If you plan it well, you can create custom columns to give you an extra tool to work with. Some people will create a custom column with an "X" or some other marker to show whether a shot has been used. One way to use sifting powers of the MC/AX is to change the search criteria to "match exactly" and have it search for a totally blank space in a custom column. This way, any shot that has not been marked is called up. The Media Tool is the only way to search for media that is online across bins, and your custom columns show up there too.

## MEDIA RELATIVES

Another way to find out if a shot has been used within a project or sequence is to use the "Find Media Relatives" menu choice. In any bin, including the Media Tool, you can choose "Find Media Relatives" in the pull-down menu in the bottom left-hand corner of the bin to find media that have clips in common. There may be subclips and master clips that share media so they are all considered media relatives.

This is the best way to search across bins to clear unneeded shots, other than the consolidation of a specific sequence and then deleting as part of the consolidation process. Earlier, we discussed what happens when you render effects and then change your mind. The effect becomes unrendered, but stays on the drive. You have a choice whenever you open up the Media Tool to show master clips, precomputes, or individual media files for all the projects on the drive or just this project. I like to show "All Projects" because many times the bins I am working with came from another project and I will not see that media if I choose only to see my project's media. Just because you open a bin in a project does not link all the master clips and media files to that project.

The most useful way to use media relatives is with sequences. First, open the Media Tool. Then you must open all your relevant sequence bins for this to be accurate and then put all the sequences that you want to keep working with into one bin. Highlight the sequences and then choose "Find Media Relatives" from that bin. The MC/AX searches all open bins and the Media Tool, and highlights all the clips, subclips, precomputes, and media that you will need to keep. You then go to the Media Tool and reverse the selection, another choice in the pull-down menu for that bin, and delete. Keep in mind that "Find Media Relatives" does not unhighlight something, so before you begin this operation, make sure nothing in the Media Tool is already selected. Also, give everything selected for deletion one final look, this is the last time you will see these files and you don't want to trust anything or anyone except yourself at this stage—there is no undo. Don't do this when you are tired! Wait until the morning and do it first thing if you can.

If you choose to show precomputes in the Media Tool, you see all the rendered effects for a project or on your drives. You can use the above method,

**Figure 4.5** Media Relatives

"Find Media Relatives," to track down the precomputes used for the sequences you've selected and then, remembering to reverse the selection, delete the unneeded precomputes. This process can be done everyday if you are working on intensive effects sequences, like a promotions department, but it doesn't need to be done unless you really need the space or memory requirements are climbing into the hundreds of thousands of objects. There is always the chance you might delete something you want to keep, so be careful.

Occasionally you will want to find exactly where a specific file is and for some reason, Consolidate, Media Mover, or the Media Tool are not sufficient to perform the task you need. In version 7.0 of the Media Composer you can use a new function called "Reveal File," which will go to the Finder level and highlight the media file associated with a specific master clip. Earlier versions of the MC/AX provide another technique that uses the third choice of the Media Tool: the individual media files. You don't need to see these very often but there is one very useful procedure. You can show clips and individual media files, highlight the master clip, and choose "Find Media Relatives." The individual media files associated with that master clip are highlighted. There may be up to nine media files associated with any master clip: one video and eight audio tracks that were originally digitized with it. Choose the track you are interested in and click on the name, not the icon, of the individual media file. Use the Macintosh copy function (Command-C) to load the name into the Macintosh clipboard and then go to the Finder level and use the "Find File" (Command -F). Then paste the long media file name into the Find File dialog box and search for it. This may be the best way to discover duplicate files and eliminate them.

## LOCK ITEMS IN BIN

If I have achieved my goal of making you slightly paranoid about deleting material during a project, then you will approach this task with the proper amount of stress. There is a great way to relieve some of this stress that does not require a doctor's prescription. A few years ago, while spending a week at a major network broadcast facility observing their operations, it was pointed out that that there was a need for someone who knew nothing about the individual projects being edited to come in late in the day and clear drive space for the next morning. On longer jobs, this person would be familiar with the exact needs of the editor. In this case, they were working on as many as five or more projects a day and at the end of the day, some material needed to stay, material that had been grabbed from their router and therefore had no timecode. But the majority of the material needed to be deleted. This is where the idea for "Lock Items in Bin" originated for Media Composer.

It was supposed to work like this: As the editors are gathering shots and using them, they decide that they need some shots for the continuing story tomorrow and duplicate the master clips (not the media). They drag the duplicate master clips into a bin named "Stock Footage." At intervals during the day, the editor goes to the stock footage bin, selects all, (Command-A) and chooses Lock Items in Bin from the Bin pull-down menu. A lock icon appears in the Lock heading that is displayed as part of the view for that bin. The file is also locked on the Finder level so that if some intrepid assistant decides to throw everything away by dragging the media file folders to the trash, they see a warning that certain items are locked and cannot be thrown away. Any good assistant, however, knows that if you hold down the Option key when emptying the trash that you can throw almost anything away and cheerfully ignore these pesky warnings.

A happy side effect of the Lock Items in Bin command is that it can also be applied to sequences. Subsequent copies of that sequence are also locked automatically when the sequence is duplicated. This helps avoid the all too common problem when people have mixed clips and sequences in the same bin and they decide to delete files. They select all and delete and then—hey, where did my sequence go? Certain types of objects are automatically selected as choices to be deleted when you Select All in a bin. In the latest versions of the MC/AX, the sequence is automatically not checked for deletion if it is mixed in a bin with clips. Select all with a bin of only sequences and they will indeed be all checked for deletion. Deleting a sequence by accident was usually someone's first visit to the Attic, only to recall the version of the sequence they worked on fifteen minutes ago! Avid Xpress does not have the Lock Items in Bin command. If you bring a bin full of locked sequences over from the Media Composer to the Avid Xpress, you can work on them just fine, but you will not be able to delete either sequences or copies made from those sequences. You can put everything into a new bin, close the bin, and then throw the bin away. Unlock clips and sequences before you bring them over to the Avid Xpress from the MC to avoid this potential frustration.

There are ways to outsmart the Lock Items in Bin command if you are determined. If you duplicate the locked clip and then unlock the duplicate then, because both clips link to the same media, you can delete the media when you delete the second clip. You can also unlock the media at the Finder level and throw it away without even opening the Media Composer application. But you can't delete the media until someone, somewhere unlocks it. Since there are really no permissions required to operate the Macintosh Finder as in the SGI or Windows NT, it is really the honor system for media management.

## CHANGING THE PROJECT NAME

In Media Composer there is another way to organize shots for easy media management: change a master clip's project name. There can be confusion about exactly which project clips are really associated with. This is because it is so easy to borrow clips from another project that you may not even know that certain shots are from another project unless you choose to show that information in your bin headings. This is why I create a custom bin view just for media management. It shows project, locked status, creation date, tracks, AVR, disk, and any other customized headings that relate to media management. If the clips have been borrowed from another project and brought into the new project, even if you redigitize them, they retain their connection to the project in which they were logged. Project affiliation also makes a big difference when it comes to redigitizing at a higher AVR because all the shots from one tape are digitized at once. If you are trying to redigitize from Tape 001 in Project X and the clip is really from Project Y, the software considers these two different tapes and asks for Tape 001 twice.

There are several steps to changing project affiliation. The first thing to keep in mind is, as mentioned in Chapter 2, project name is part of the tape name. The trick to changing the project name is to change the tape name to a tape name from the right project. In versions 7.0/2.0 all tapes have a Project column in the Select Tapes window. This allows you to see which project each tape is from. In Capture Mode, whenever you show the list of tapes that have been digitized and you check "Show other project's tapes," you might see multiple Tapes 001 from other projects. You occasionally need to use the Scan for Tapes button in versions 7.0/2.0 to make sure the system has looked at all the tapes digitized and used on the system.

Open the new project and, within that project, open the bin from the old project that contains the shots that you want to permanently change to a new project name. Create a new bin and copy the clips into the new bin. Select all the clips. Go to the Modify command under the Clip menu and modify the source name. If the tape name you want has not been used before in this project, you can create a new tape name. You can choose an existing name if you made an earlier mistake and need to rename these clips so they all come from an existing tape in the new project. As you finish with the modification process you get a series of warnings that are important if you are working with key numbers in a film project since the new clips need them re-entered after you modify the tape name. A film project will not allow you to change a source name of a clip without unlinking first. Otherwise, don't worry and just check OK.

Now the clips are associated with the new project as far as the Media Tool and the headings in the bin are concerned. The change has not really affected the actual media file at this stage, just the master clip in the project. This is not enough for Media Mover to recognize there has been a change since it only looks at the media file and not the project. The project information about a media file is actually recorded in the media object (MOB) ID (it has nothing to do with gangsters). This MOB ID is attached to the media file when it is digitized and the media file itself must now be changed to update this information. Here is another use for Consolidate. You need the space on the drives to copy the files you want to change. Select them all and Consolidate with the following options checked: have the old name link to the new media, do not skip over the media if it is already on the drive, and delete the old media when you are done. Now you have copied the media and the new copies have the new project name. If you want two sets of the media, one in one project and one in the other, do not delete the original media as you Consolidate.

## BACKING UP MEDIA

Now that you have eliminated the unneeded material by deleting, whether it is digitized media files or precomputes, you can now evaluate the choices available for backing up. As very large media drives become a reality, there may be a room full of gigantic backup drives in your future. Clearly, this is the fastest way to copy media and, if it is done over a very fast network like Fibre Channel, those drives could be anywhere in the building or, eventually, the world. Perhaps there is a future business for someone to set up thousands of drives in a warehouse somewhere in the desert and have people rent out the space for backing up over the Internet?

But that day is a few years away and for now, you probably have to deal with digital linear tapes (DLTs) or another form of archival tape backup. Until recently, 8-mm drives were considered along with DLTs but the size of the 8-mm capacities has not grown as fast as the size of the DLTs, although they are cheaper. Looking at the advertisements for different kinds of backup products, it may seem that an 8-mm drive or a non-autoloading DLT is more than adequate for your storage needs and you can pocket the extra money and use it for that industrial cappuccino machine. This is because most advertisements feature the amount of storage they are capable of when the data is compressed. But you have already compressed. You are working with Motion JPEG and Avid's compression schemes have already squeezed the heck out of it. These files are not getting any smaller, which means you should always choose to avoid any further compression as an option during your backup session. Further compression takes extra time and adds a layer of unneeded complexity to already very large and complex files. You may be courting data corruption and that's not a cheap date!

The linear part of the DLT is what most people object to when using them, plus the fact that they can be rather slow to back up and just as slow to restore. Just like hard drives, they too get faster all the time. The fastest DLT that Avid sells backs up at the blinding speed of 5 megabytes per second and it can back up or restore 18 gigabytes in an hour.

Many people use DLTs to archive material that has no timecode. If they were to delete and redigitize or reimport non-timecoded material they would have to re-edit it in by eye. If the material is graphics, then they may have many layers to step in to re-edit and potentially some very complex sequences to modify. I still recommend dubbing non-timecoded material to a timecoded format but sometimes, especially for fast turn-around projects or lots of graphics, this is just not practical. You also lose the quality benefit of keeping the images completely digital and uncompressed until they are imported into the MC/AX, unless your timecoded medium is also uncompressed. Dubbing non-timecoded material to tape with timecode is still the simpler and more reliable way to get your material back on the drives and get it to link to your sequence.

Many times it is actually faster to redigitize a shot or two than rely on a slow and cumbersome restoration process and this is a reason why people don't back up their media as much as their project information. The exceptions to this would include video that has been carefully color corrected during digitizing and the color correction information has not been saved. Also, there are many instances where the original tapes become unavailable either because they are sent to another facility or, working without a net, they are recorded over. Again, the question is not if your hard drives die; it is when. What is truly important is not how easy it is to back up but how easy it is to restore.

The professionally paranoid know that something is really out to get them and you can never be too prepared. For people like this, it is usually a good idea to do a full backup of all material after the first day of digitizing. From that point on incremental, or normal backups can be scheduled every night to store only the material that has been added throughout the day. You may want to even consider backing up over a fast network so that you can back up all the editing systems from a central location. With the size of storage on even one machine easily overwhelming the capabilities of a non-autoloading DLT, this may be just wishful thinking. You may need a DLT for each of your important systems. Figure out how much it would cost you to lose that information for even four hours and then compare it to the cost of a DLT.

A new choice for backup software that has begun to ship with DLTs when you buy them from Avid is MezzoArchiver from Mezzo Technology. This software backs up in the background throughout the day. You can identify all the material that should be archived by just dragging the project folder to the graphical interface of MezzoArchiver and leave the DLT connected to work in the background. You will find at the end of the session that only the very last material digitized or rendered needs to be backed up before you can go home. MezzoArchiver also allows you to target specific clips for restoration instead of the entire project. Since MezzoArchiver can identify all the material attached to a project, search for it on all the drives connected, and back up incrementally, you are operating with a safety net. For an emergency, like when a single non-timecoded music file has been deleted, restoration will be a fast and painless process.

The only reason that a DLT is not a foolproof way to back up media files for archival purposes is that over time, there may be a significant change in the structure of media files. This may occur because of a software change, as in versions 7.0/2.0, when the structure for the media files changed from MSM to MFM to be compatible with NewsCutter and the EditCam and in the process became OMFI native. MSM ensures that all media files are "server ready" no matter which edit system you are using. If you upgrade to that version of the software, then all of your media files must either be converted to the new format or redigitized. When Avid made the switch from the NuVista+ video board to the Avid Broadcast Video Board, there was a change in the size of the frame as well as a jump in quality. You had to convert the NuVista+ media, which was digitized at 640×480 square pixels to the 720×486 non-square pixels. No pixel-based images are improved by expanding and distorting. The way to take advantage of the new technology was to fall back to the old technology and redigitize from tape. In the short run, or if you are not planning to upgrade any time soon, the DLT is your best answer because it allows you to keep entire projects with all the rendered precomputes and non-timecoded material all in one easy-to-restore package that takes very little space on a shelf.

## OTHER OPTIONS

There are other options that don't include backing up everything—just those items that can't be redigitized. The slower, optical drives are not supported by Avid with the newest software for playing or digitizing, even the Avid Optical drives. Many people are considering the new removable 1 gigabyte cartridges like the Jaz Drive by Iomega because they are as fast as a slow hard drive. Some people are even brave enough to play back audio from these drives (not supported)! You can improve performance of these drives by buying two and striping them (really not supported). If you misplace one of the striped cartridges, the one you have is not readable.

The removable tape drives are far more expensive per gigabyte than the DLTs, but then you need to be choosier about which media actually gets backed up to these smaller, faster, removable backup drives. You must be able to put all non-timecoded material on one partition so that it is only that partition which is copied to the backup drive.

## BACKING UP PROJECTS AND USER SETTINGS

After all these methods of backing up media are considered, don't forget to back up your projects too! This is the most important thing you back up on the MC/AX and is all you need to recreate any project, except for non-timecoded material and rendering. You may want to back the project up to several places as well as putting it on the DLT.

There is the common question of backing up User Settings. You should be able to reproduce User Settings in about 5 minutes, but some people resist it like a trip to the dentist. There is no way to lock your User Settings to keep unauthorized folks from making a few "improvements" or just accidentally changing them, so it is a good idea to back them up somewhere safe and hidden. Many freelancers carry their User Settings on a floppy and other, more intrepid instructors or field support technicians carry their own small, fast hard drive with everything they could possibly need.

When software versions change, especially substantial version changes, you should always remake your User Settings. I know this sounds tedious, but recreating them solves many unusual and unpredictable problems, especially if your customized keyboard is complex. You may be trying to access menus that have been moved or deleted! Or there may be more subtle problems that don't immediately appear to relate to User Settings. One of the first things Customer Support asks you to do if you are getting unusual behavior, like a common feature suddenly not working, is to recreate your project, bins, and User Settings. Having a backup of your User Settings may be enough to fix the problem, rather than having to recreate them from scratch, if the software version has not changed recently.

## NETWORKS

Any good administrator must evaluate how a network can improve their operation. One of the smartest broadcast engineers I ever met told me he was studying networks at night school. Very much like plastics was the secret answer in the 1960s, networks may be the not so secret answer in the next decade. Networks, as they exist today, are slow, but just like hard drives and DLTs, they are getting faster all the time.

The question for you is, "How much is it worth to have a network?" and plan your investment accordingly. As you may suspect, the more money you invest, the faster you can go. Whatever you do, do not rely on the old standby LocalTalk. This is generally incompatible with the MC/AX and has been known to cause some pretty serious problems during digitizing and digital cuts. Besides, who wants their sessions interrupted with e-mail notices? The only thing you might consider using LocalTalk for is tying into a printer, but you will have to restart the computer to activate it. Over time this will prove to be more awkward than useful and eventually you must get a printer that does not need LocalTalk, uses a LocalTalk converter, or connects directly to the serial port of the Macintosh.

Better to at least look at Ethernet. It may not be the fastest network choice out there, but it sure is easy. All systems currently shipping from Avid on the Mac platform are Ethernet ready. Plug and play and all those printer problems go away if your printer is Ethernet ready, too. You still probably don't want that e-mail icon flashing in the right-hand corner though. Any time you are asking your system to be interrupted by incoming signals, you are asking for some kind of error message during CPU-intensive tasks like digitizing, digital cut, or rendering.

Ethernet can be easily expanded by the addition of an Ethernet hub. For very little money you can tie all your editing and logging systems together on one local network. The best reason to use plain vanilla Ethernet, technically known as 10 Base-T, is to share project information and small graphics. It is also a good way to set yourself up to back up just the contents of the internal hard drives with project, bin, and script information every night. You can use it to send the occasional media file but there are better options for that which we will cover later in this chapter.

The 10 Base-T refers to the size of the wire connecting all the systems. Think of it as a thin, but adequate pipeline that maxes out at a total capacity of 10 megabits per second. Don't expect to be blown away by the speed, but you will appreciate the simplicity. For serious network use, consider adding some more money and another circuit board to the equation. The next step up is Fast Ethernet with 100 Base-T and, as you have probably guessed, it is faster and uses a thicker cable for the connections. It also requires a small addition to your CPU so you need an extra slot inside your computer for a circuit board that will allow you to accelerate the Ethernet. This allows you to consider larger graphics, digitized audio, and the occasional video media file sent over the network. Many people use this to tie into their digital audio workstation.

## MEDIASHARE

MediaShare uses a special piece of hardware, a SCSI switching device, that Avid supports for three editing systems to connect to the same drives at a distance of up to a total of 150 feet (75 feet to MediaShare and another 75 feet to the drives). Editor A can be cutting using material from the group of drive towers while the assistant is digitizing on another system to the same drive tower. Or, Editor B can be working on a second system while Editor C can be outputting a digital cut. There are strict limitations with SCSI MediaShare because it is still a SCSI system and not a new kind of network. SCSI MediaShare is limited to three drive towers and can reliably work with all single-field AVRs. Although many people push these limitations and can work beyond them, things don't work quite as smoothly.

There are also administration limitations to MediaShare because it relies on a piece of software called MS Scanner. Only one system, the administrator, can have the privileges to do media management on all the drives and this one person can decide the privileges for the entire group. If you do not work at the administrative system, you cannot write (digitize, render, import, etc.) to certain drives unless you have permission. Changing administration privileges or performing drive maintenance requires shutting down all the systems. With lots of media online, several towers full of low-resolution clips, restarting can take around twenty minutes.

If you have a traditional feature film setup of one or two editors and an assistant and this setup will not change during the project, then SCSI MediaShare may be adequate. If you have a fast-changing situation, it may not give you enough flexibility. It will never be fast enough for online quality material and will never go much beyond the current capabilities.

## FIBRECHANNEL

A Fibrechannel-Arbitrated Loop is so fast that you will find yourself not copying files across this network, but actually using media files from a central group of drives that everyone shares. In the past, this functionality has only been available when using MediaShare and now Avid has released MediaShare FC. Fibrechannel makes it possible to send large media files at speeds that will equal or exceed the amount of time necessary to play these shots back in real time. Theoretical speeds of transfer of 100 MB per second will allow five simultaneous users at AVR 75 or ten users at AVR 8s. There is no server required, no IDs, over 100 devices per loop, and up to 30 meters between drives or workstations. This is the network to get if you are really serious about sharing or exchanging media files.

## AVIDNET

Fortunately, Avid has made it even easier and faster to use these kinds of networks by creating their own file transfer protocol called AvidNet. This means that Avid uses the wire connections, but designs their own way of sending the information through the network. With AvidNet you can send material only between Avid systems that use the AvidNet protocol. It allows transfer of material faster than standard Ethernet by making up for inherent compromises in the Ethernet protocols.

AvidNet is also easy to use since it works from inside the editing application. What could be easier than an inbox and an outbox? The user at the first station copies the material to be sent and puts it in an outbox on the system then alerts system 2 that the media is ready. System 2 receives the media files automatically. This is known as a *push system* because System 1 pushes the material onto System 2.

The ease of use really is evident when you want to just consolidate a sequence. There is no need to consolidate first and then transfer because, on a Macintosh based system, AvidNet figures out what is required for the sequence, places it in the outbox and consolidates simultaneously. This is the best way to send just what is needed for another editor to continue to work on a sequence at another station.

## SERVERS

Any administrator should also consider setting up a server. Simply, a server is a stand-alone Macintosh fast enough to handle large transfers of information and it is connected as a central storage place for resources important to everyone working on the project. It may be a place to store the final version of any graphic, logo, or script so the chances of someone using an older or unapproved version of something are reduced. Many times people end up using the wrong version of something because it was changed on one system and not copied to all the other systems. There should also be a good policy to enforce and maintain a level of compliance.

On one project we were so concerned about individual editors changing the tape named bins, that everyone was asked (repeatedly) to digitize and edit from bins that were on the server, but not on their system. In other words, they would access all bins over the network. The server icon would appear on their desktop like another drive and they could open any bin stored there just as easily as any bin actually on their own system. If they wanted to customize or modify a bin they were forced to copy the bin to their individual system and from that time forward, it was their own responsibility to keep track of that bin and name it for themselves. This maintained a pristine copy of every original bin in the project in a place where all editors could get it at any time. The downside was if the server crashed, they crashed too. Unlikely but possible and they would still have their own Attic.

A server is also a good place from which to base a facility-wide backup or just to connect a Zip drive to back up central bins every so often without disturbing anyone else. It is also a good place to connect a printer for everyone to use. An assistant can be printing out bins non-stop and never interrupt any editing or logging, even though the server could conceivably be used as a logging station in a pinch. The server can be a good place to do database searches with a program like Retrieve-It! and MediaLog or to make EDLs with EDL Manager. You don't need a computer full of fancy circuit boards to open a bin! Besides, what else are you going to do with that old Power Quadra 950?

## SNEAKER NET

Finally the last, cheapest, and worst choice for networking: the Sneaker Net. If you have invested in the MediaDocks this is not so bad. You just shuttle the media from room to room by hand (sneakers are optional). It means stopping work on two systems until the transfer is complete. If the goal is just to take everything to another location, it is simple and easy. Just don't trip! (Shoelaces should be tied!) If at all possible, don't fall into the common practice of disconnecting standard, fixed drives and moving them from room to room. You may get away with this for awhile but eventually you end up with bent pins, kinked cables, and a few error messages. Make sure the person doing the Sneaker Net transfer knows everything there is to know about SCSI configurations; this may end up being you!

## RELINKING

All master clips, subclips, sequences, and graphics must link to media in order to play. When you log a shot into a bin you create a text file. When you digitize the shot you create a media file on the media drive. The relationship between the master clip and the media file is considered a *link*. If the master clip becomes *unlinked* from the media file, it is considered offline. The media may be on the media drive but if the master clip is unlinked, it is considered offline and unavailable. In order to link you must highlight the clip and choose Relink from the Clip pull-down menu. The ability to link, relink, and unlink is very important to sophisticated media management. We will revisit these concepts over and over again in this handbook. Here are ten rules that govern this complex and confusing behavior.

### Rule 1

All linking between a master clip and a media file is based on identical tape name and timecode with media in the 5.*x*, 6.*x*, or OMFIMediaFiles folder. This means that when you have a master clip, subclip, sequence, or graphic that is offline, or not connected to any other media, and you know that the media is on the drive in the MediaFiles folder, you must specifically instruct the new media to relink. Highlight the master clip and choose Relink from the Clip menu. Your MC/AX searches the active MediaFiles folder for media that uses the same tape name and timecode.

### Rule 2

Just because the tape name looks the same doesn't mean it is the same tape. There can be only one tape logged per project with a particular name. Every time you add a tape as New in the digitize tool, you are creating a unique tape that is associated with that project. You can use two different tape 001s in the same project, but they are logged in separate projects and are always considered different tapes. In pre-7.0/2.0 versions if you uncheck "Only show this project's tapes" in the tape naming dialog window, you see the project name attached to the tape name. Tape 001 becomes Project X:001. In versions 7.0/2.0 the project name is always displayed with the tape name in the tape naming dialog window. To reduce confusion, all tapes for a project should be logged in the same project or in another project with the exact same name.

## Rule 3

A master clip cannot link to digitized media that is even one frame different from the master clip's start and end times, even though it has common timecode and tape name.

## Rule 4

A subclip will relink to media that is longer than the subclip, even if it is exactly the same length as a master clip which will not relink. Subclips are programmed to expect to link to more media than the subclip start and end times.

## Rule 5

A sequence may relink when the individual master clips that are contained in it will not. Think of a sequence as many subclips.

## Rule 6

A sequence or a subclip will not relink to a media file that is shorter than the media it needs.

## Rule 7

The Relink Master Clip(s) checkbox in the Relink dialog box must only be used with master clips.

## Rule 8

In the Relink dialog box, the Relink to Selected checkbox can only be used when you have highlighted specific master clips that you want to relink to a sequence. This option only works within a single bin. It is primarily used to link a sequence to media at another AVR. To relink to clips from the Media Tool, the clips must first be dragged into the same bin with the sequence and everything must be highlighted. Do not check this function unless you are specifically linking "to selected" or nothing will happen.

## Rule 9

If all of your efforts fail to produce a relinking, make sure you are working with the version of the bin that was actually digitized (i.e., the same project). If this still does not work, you will have to rebuild the media databases and desktop file. Your drives may be too full to correctly rebuild the media databases and you should delete unneeded precomputes or media to create space.

## Rule 10

Sometimes a link must be deliberately broken using Unlink. In the Clip menu, highlight the desired clips, hold down Shift-Control and Relink becomes Unlink. You can then modify duration of the clip in the MC/AX but you must redigitize all the clips. Do not unlink media that has no timecode and do not modify the master clip because you will be unable to batch redigitize.

# USE COMMON SENSE

So while there is a lot to be in charge of in administering an Avid system, much of it is common sense and taking advantage of existing computer peripherals and software that make your job easier. Whatever policy you decide on, make sure it is followed uniformly and make sure that all members of your staff are educated, not only on the correct procedures, but just a little troubleshooting, so they can deal with those questions themselves during the overnight shift. If set up right, you will significantly reduce downtime and make it easier to diagnose and solve technical problems and missing data.

# 5

# Long Format and Improving System Performance

Whenever you push a system to the limits of its capabilities, different rules of operation may apply. All of the functionality described in this book should operate correctly no matter how many drives or how long the sequence. But any time you work with extremely large files on a computer, you want to minimize the wait times involved with the extra work all the hardware must do to keep up. As with all programs that give you an advanced level of flexibility, the Avid allows you to work only on the specific parts of the project you want to change. If you are working on a sound mix you may want to minimize the video or vice versa. This chapter describes various techniques developed specifically to increase performance and to cope with situations that occur only when working on longer format pieces.

Any sequence longer than a few minutes that requires more than just a few drives for the media could be considered long format. Only as the number of drives, master clips, and the length of the sequences creep up will you experience any benefit from many of these procedures. But many of these techniques will speed up the response of the systems no matter how long the project.

If you experience a sudden slowdown in performance, you may be experiencing a simple technical problem which can be troubleshot and solved quickly. Generally, a sudden slowdown occurs if lots of very large bins are being opened and closed or other pieces of software have been launched and quit throughout the day. This is just normal RAM fragmentation in the Macintosh and can be solved easily by shutting down the computer and restarting. Most people do this before the beginning of a shift and some, working with lots of media and bins, also shut down at lunch and then restart. Dismounting drives that hold unneeded media, by dragging the icons to the trash (or highlighting them and using Command-Y), can instantly reduce the amount of objects the system needs to account for. You may want to rebuild the desktop and the media databases as well. Details on those procedures are covered in Chapter 9.

The absolute first thing that must be considered before starting a long project is how much RAM is in the system. Over 150 megabytes of RAM is not too much. The more objects or digitized master clips, precomputes, sequences, subclips, etc., the more RAM you need to handle them. As I have mentioned before, any cost for RAM more than pays for itself as the project gets longer and more complex. The only bad side effect of having so much RAM is that the startup process of the Macintosh takes considerably longer; sometimes a few minutes before you even see the happy Mac icon. All of that RAM must be initialized and that takes time.

## BUILDING THE PIPES

Every time you press Play on the MC/AX there is a pre-loading process where all media in the sequence must be scanned for and analyzed. If there are mismatched AVRs in monitored tracks or mismatched sample rates in the monitored audio, you may get an error message and the sequence will not play. You may see the video slate "Wrong Format" when trying to play any AVR that could not mix with the AVR of the very first shot of the sequence.

The benefit of this pre-loading or "building the pipes" is the ability to play the entire sequence backward or forward at any time. The drawback is that the system is working very hard to provide access to all your media even though you may only want to see the next 30 seconds. When the sequence gets quite long, there may be a noticeable delay after pressing the Play button.

There are three methods to reduce this delay. The first method is to work with shorter sequences. This has many benefits, including not having to worry about sync 20 minutes down the timeline when you just want to make a few quick changes to the opening montage. Focusing on a specific trouble section and massaging it without worrying about affecting the rest of the sequence cannot help but speed up the editing process, even disregarding technical playback limitations. Many projects lend themselves to being broken into smaller pieces anyway. Films break easily between scenes, episodics break between acts and corporate programs and documentaries may break at location or subject changes. When it becomes necessary to view the entire project at once, it is a simple process to Shift-click all the smaller sequences and drag them to the record monitor where they instantly become a composite master. Perhaps a fade to black or dissolve needs to be added between these segments, but otherwise you have a quick, rough assembly of the entire show.

After the approval viewing or the digital cut is over, go back to the individual segments to make the changes and archive the composite master. There is nothing worse than making changes on a short, individual segment but then showing the older, composite master sequence without those changes!

The second method is to only "Play to Out" or Option-Play. The system only builds the pipes up to the marked-out point. This approach will be necessary at a later point in the project, after all the segments have been combined or if there is no way to break the sequence into smaller pieces. If you get into the habit of quickly jumping down the timeline a minute or so, marking an out and then using Option-Play, you get a faster response from the system.

## PLAYLENGTH

The third method to reduce playback delay requires typing a command into a window under Tools called the Console. The Console is powerful but potentially dangerous because it is designed primarily for the programmers to debug the program or run various tests. Here is the obligatory warning: Any command typed into the Console is completely unsupported by official Avid policy. Under the wrong conditions you can permanently change the way your software operates and you may have to reload it from the original disks to get it to behave correctly again. Don't mess around with the Console unless you know what you are doing!

Despite all these warnings, there is one very useful command that has worked well in the Console for years called *PlayLength*. If you type "PlayLength 2" and Return, for instance, for the rest of the session (until you relaunch the software) you limit the building of the pipes to two minutes total. This means that when you press Play, the MC/AX only pre-loads one minute of media before the present blue bar position in the timeline and one minute past it. When it reaches the end of the one minute, it stops playing and you need to press Play again. Because the pre-load is so much shorter, the response time of the system for much of its operation is noticeably snappier. PlayLength does not affect the Digital Cut. This prevents a possible embarrassing situation when you go to tape, or as some very brave souls have tried, straight to air. To return the system back to normal, go back to the Console and type "PlayLength 0" and Return. Building the pipes has been noticeably improved in versions 7.0/2.0 and it is now a user selectable button in Media Composer.

**Figure 5.1** Specifying PlayLength

## RENDER ON THE FLY

Other ways to improve performance is to utilize techniques to reduce the amount of media that must be retrieved, played, or rendered. There is a menu choice under the Special pull-down menu called Render on the Fly. Render on the Fly is a temporary render of one frame performed in the software when you place the blue position bar over the effect in the sequence. Nothing actually goes to a media drive. With Render on the Fly turned off, when the position bar stops over an unrendered effect, the system does not try to composite that single frame for viewing. The system shows the highest numbered track that does not contain an effect. This means you can scroll through the timeline and stop in the middle of an unrendered effects sequence without the delay involved in previewing. You can move quickly to perform tasks that do not require seeing these composites including audio mixing, trimming to marks based on audio, and moving segments around the timeline.

Editors who make use of turning off Render on the Fly on a regular basis end up mapping it to a function key. The only drawback to disabling Render on the Fly, besides not being able to see every layer, is that some objects like titles or imported matte keys are considered "effects with source." This means that they must be rendered on the fly to be seen. If Render on the Fly is disabled, you only see black when loading a title into the source window. Ironically, you can view them as soon as you hit Play because then they are rendered in real time. Leaving Render on the Fly disabled has resulted in a few calls to Avid Support, so remember to turn it back on before handing the system over to the next editor.

There is also a checkbox for enabling Render on the Fly in the trim mode under the Trim Setting. It defaults to being off, but once turned on, it allows you to see multiple, unrendered layers and effects while in the trim mode. This slows down trimming considerably, but may be necessary to make content-based trims with those layers in view. If you find yourself needing this on a regular basis, make two separate Trim Settings and switch between them in the Project window.

## MOVING THE VIDEO MONITOR

Another way to improve performance while working with effects is to move the video monitor icon to a lower track. This means that since you are not monitoring those tracks, the media does not need to be retrieved and played. Again, you can move quickly through a multi-layered sequence without waiting for unneeded images to be recalled. To move even faster, turn the video monitor off completely. Sound mixing and rearranging of large chunks of the sequence will speed up remarkably. This is also a good way to isolate whether playing back a video track is causing a particular error. If you stop monitoring the video track and the playback problem disappears, you have isolated the problem to that track.

## RENDERING FOR SPEED

By occasionally rendering real-time effects during a break, you can make it easier for the system to cope with playing many short clips. The system is always trying to display all real-time effects by doing the compositing in the CPU. Playing back a precompute, a rendered effect file, takes less effort and can result in faster response time. If you are having problems playing a particular section of fast, short cuts, you can apply a submaster to the track above and render it. This puts all the short media files into one rendered precompute so the system only needs to play back one single file when you monitor the track with the submaster. When you make the EDL, you avoid the video track with the submaster and still generate a clean, accurate list.

## VIDEO MIXDOWN FOR SPEED

If you don't need an EDL, you can create a video mixdown of the section. A video mixdown combines all the separate media files and precomputes and makes them one new media file and a new master clip. This also plays back very quickly with the added benefit that, if the segment has several layers, those media files required for the layers can be completely eliminated. If all the effects in a sequence have been rendered, then making a video mixdown is as fast as copying a file from one part of the drive to another, an insignificant amount of time compared to the gains of faster access time.

Even after you render an effect, the layers below the top track continue to cause media to be retrieved in case it is needed. With video mixdown, there is just one layer, one media file, and subsequently, faster retrieval time. Don't use a video mixdown if you need an EDL or you need to matchframe back to an original source in the segment. A video mixdown has no timecode and retains no links with the original media; it is purely a convenience for simplifying a complex segment. Always make a duplicate of the sequence before cutting a video mixdown over the original layers so you'll have a non-mixdown version of the sequence to come back to for revisions.

## MINIMIZE TIMELINE REDRAW TIME

There are ways to minimize the screen redraw time of the timeline when there are many icons that need to be constantly updated. You can go into the Timeline fast menu or Timeline User Settings and start disabling the way that various icons are displayed or turn them off completely. By turning off effect icons and dissolve icons, you lose that visual information, but if you are positioned over the clip, you can still see the effect in the record window. Effect Contents can be disabled so that if there are nested effects, they only show up as the outside effect without trying to show all the layers inside. If you have not seen this setting at work, it is because it is only visible when you are zoomed in closely on the timeline and have enlarged the height of that track (Command-L). Never work with the audio waveforms displayed unless you specifically need them. This can be better utilized in version 7.0 where there is a Timeline User Setting that shows waveforms only between mark inpoints and outpoints. Any of the modes that require images to be drawn in the timeline should be disabled, like Heads View or the Film track, and in general, any display you can work without temporarily should be turned off. I always leave sync breaks on, but I might turn them off for audio when I am editing primarily video. You should save a Timeline view with all these display features turned off so you can get to it quickly when you really want to move fast.

An instance where redrawing images can slow performance occurs when using the Segment editing arrows. As soon as you begin to drag a segment in the timeline, the four screens showing all the transition frames being affected must be drawn on the screen. When moving especially fast this redraw can delay the system enough so that it perceives a click and drag as a double-click. The system waits for you to release the mouse so it can display nesting, the command it has received from a double-click over an effect in the Segment mode. It does not allow you to drag at all. In earlier versions, this four-screen monitor can be

**Figure 5.2** Four-Screen Display

turned off for the session and in version 7.0 of Media Composer, it can be disabled as part of a User Setting under the new Timeline setting (Show Segment Drag Quads). The old fashioned way (version 6.5.x and earlier) to disable the four screens in segment mode is to click in the Timeline with a Segment Arrow and, while still holding down the mouse, hold down the Shift button and drag the segment slightly. You hear an error beep and the four screens are off until you perform this procedure again or restart the system.

With the new Timeline User Setting in Media Composer there is also a command to turn off the Segment Arrow, double-click, that displays nesting. These are both welcome additions to the speed freaks out there who are always working a few steps faster than the system can display.

Working with long projects means trying to squeeze all the extra speed out of your system. It means stripping down the features to the bare necessities to minimize screen redraw time and wasted time accessing images and effects you don't need at that moment. In the end, you are working with the most flexible, customizable system there is for media management and maximizing the use of the drives and CPU. No other system comes close.

# 6

# Importing and Exporting

A major benefit of working on a desktop system for video is the ability to import from and export to other third-party programs. Bringing in graphics from new and different sources is not only convenient and fast, it all but eliminates the traditional copy-stand work that was done in many online suites.

Importing and exporting with the MC/AX depends on the end result. Do you want the highest quality? You must follow some basic requirements for format, size, and resolution. Do you want to play the material on a CD-ROM? You must make some quality tradeoffs to get a decent throughput of the images and audio on consumer computers. You must, above all, educate your subcontractors and graphics department (even if that is just you!) and find a consistent procedure for acquiring the type of graphics and animation you demand for your clients and productions.

There are some basic issues that should be understood when working with computers, graphics, and video. Because of the way computers have developed, their reliance on RGB color for their screens and memory, and the way video developed, beginning with black and white, the two mediums have never had a particularly easy co-existence. One of the hopes for high-definition television is that some of those issues will be resolved, but the addition of another, incompatible format has rarely made things simpler.

## COLOR SPACE CONVERSION

Computers and video work with different color formats and different kinds of scanning and scan rates. Computers traditionally work in RGB color space that is easier for displaying images on an RGB computer screen and working with computer memory. RGB has a more limited span of colors to choose from, but it is very dense, with many more possible points within that range of perceptible colors. The ironic part about camera-originated video is that it begins as RGB and is then converted by the camera to a different color space called YCbCr (or YUV).

This new YCbCr color space spans beyond the range of visible colors, but has fewer points within the visible range. Thus, a conversion from YCbCr to RGB can force colors to change because a true exact match does not exist or because colors are out of the range in the new color space ("out of gamut").

The change of colors when the video camera's RGB values are captured and then converted to YCbCr as the signal is recorded to tape is a greater source of color change than anything that happens later inside the computer. The amount of color information is cut in half as you move from a 4:4:4 signal within the camera to a 4:2:2 signal on tape. So yes, there is potentially some change to your video colors during the digitize process, but you will most likely never see it on a standard video monitor. The conversion never affects monochrome values no matter how many generations you go using Avid's Macintosh video board, the ABVB. If your colors or levels are changing it is more likely you are exporting or importing into the MC/AX using the RGB levels choice instead of CCIR levels. This chapter will discuss the details of these choices later.

The Avid Broadcast Video Board (ABVB) is the video board for the Macintosh used in the MC/AX for MC version 6.0 and later and for all Avid Xpress models. It is ITU-R 601 compliant. (This regulation used to be called CCIR-601 but they changed the name to reflect another committee.) ITU-R 601 compliant means that it follows the standards for digital video. This makes things a little more complicated because most computer graphics programs do not conform to this video standard. At the same time, it opens up some interesting possibilities.

The reason it is more complicated is that the early video boards (the NuVista+ included) and prosumer video boards, although some may be of good quality, do not capture the entire range of the possible video levels. A true ITU-R 601 video signal can use a signal which is as black as possible without going into the sync section of the signal. It can also go above the brightest level that should be broadcast, giving some extra range for the highlights. The other, older design boards use a more limited range, which in NTSC is from 7.5 IRE, which is the blackest you can legally broadcast in NTSC, to exactly 100 IRE. In PAL the range can be shifted but usually goes from 0 volts to .92 volts.

Not allowing a video signal below 7.5 IRE means that you cannot use the levels all the way down to 0 IRE (superblack) for keying graphics in a luminance key. This is one of the most popular and easy-to-accomplish graphics techniques in use with video-originated graphics today and this chapter discusses it later. Many times in a broadcast situation, the last step to the production is to place a small "bug" or logo over one corner. Using superblack allows that logo to contain black in the design and still be keyed, because the level of black that is used to "key out" is at 0 IRE. This allows any 7.5 IRE normal black levels to "key in" or remain visible. If the only way to bring the graphic in to MC/AX is from video tape and there is no mask, matte or hi-con, you need to key the graphic with a luminance key.

Another characteristic of the ITU-R 601 requirements is that the frame aspect ratio is different from most graphics computer programs. With standard computer monitors, most people work with square pixel graphics with a frame size of 648 × 486 pixels. This is fine if you are going to one of the older designed, non-ITU-R 601 video boards, but the 601 standard size and the ABVB require the graphic to be 720 × 486, a non-square pixel size. If you are importing a graphic that is 648 × 486, this is not a terrible mistake, but it means that the MC/AX has to resize the graphic on import for you. This adds another step to the process and, if you have a lot of fine detail in the image, on pre-7.0/2.0 versions you may see some resizing artifacts.

The non-square pixel adjustment is not just a question of resizing, since it also affects the aspect ratio of geometric shapes by slightly distorting them. If you create a circle in a square-pixel graphics program like Photoshop and then import it into a non-square-pixel video program like the MC/AX, your circle will turn into a slight oval. It is stretched to fit the same number of pixels into a different aspect ratio. To deal with this mismatch, create the frame slightly bigger than the standard square pixel size but keeping the right aspect ratio, 720 × 540 (NTSC) or 720 × 576 (PAL), and then squeeze it to the right aspect ratio for non-square pixels by reducing one dimension. If you squeeze the image before importing, it will look correct when it is finally in the MC/AX. This chapter will cover this procedure in more detail.

The final characteristic of graphics and video that must be taken into consideration is a video field. There are two fields to every video frame since a standard television signal scans the screen twice. Whether you are PAL or NTSC, you must use interlaced scan lines—every other line is part of a different pass drawing the image on the screen. Broadcasting this signal takes less bandwidth because only half of the image is transmitted at any one time, but it complicates things when you are taking an interlaced image to a non-interlaced medium like the computer. A computer uses a progressive scan which means that one line is drawn on the screen, then the next, and then the next, on down to the bottom of the frame and then back to the top and the process starts again. If you export an interlaced frame from the MC/AX and there is some kind of horizontal motion in the frame, you see a difference between the first set of scan lines, field one, and the second set, field two. Even though they are only a 60th of a second apart, you see jagged horizontal displacement of the image every other line.

All of these mismatches can be dealt with if you follow a set of guidelines that takes one kind of image and converts it to another. Sometimes there is a loss of quality as the transition is not absolutely perfect, but hopefully you can plan around these present shortcomings when using the mismatched formats.

Most images you have seen on television up to the last few years have, at some point, passed through a camera or been laid off to videotape before being edited into a program. With the ease of basic desktop editing and the combination of graphics that go straight from one computer graphic format to a computer video format, you have a set of new problems. These graphics have too much fine detail to reproduce well in the relatively low-resolution, interlaced world of video. Consequently, they buzz and flicker and give us unpredictable results. A thin line may look fine on a progressive scan monitor, but the moment it moves to an interlaced scan medium like video, that line may be only one scanline wide. This means the line is only drawn on the screen every other field and causes a disturbing flicker. Images that originated with a video camera can never be recorded with that kind of problem. And since, in the past, most graphics were seen through video monitors as they were being created, they could be adjusted on the spot so that the design could take into consideration the limitations of what looked good on video. Colors were toned down and detail was blurred until the image was acceptable and then it was put to tape.

Until all graphic workstations can figure out how to approximate what the final product will look like after being imported into the MC/AX, then you must be able to tweak the graphics yourself. The most common adjustments are to open the graphic on the editing workstation and add a little blur to areas that are buzzing with too much detail. If done with a little skill, it will never be noticed. In fact, a very slightly blurred image looks better than one that is too sharp. With version 7.0 of Media Composer, there are painting tools that allow you to do this without any third-party programs, but they must be rendered. Another common adjustment is to lower the saturation of a particular color or, in a worst case scenario, all the colors. Again, this can be done easily and safely using the MC/AX color effect or by running all the images in a batch through a program that allows you to program macros like Equilibrium's DeBabelizer.

## IMPORTING REQUIREMENTS

There are three important factors to get right when importing images into the MC/AX. They are size, resolution, and format. If you get these three characteristics correct, then any graphic should import and be represented correctly. The latest version of the MC/AX handles many different formats, but these three characteristics must be right or the MC/AX either ignores the image or distorts it.

Size and resolution are related. For the ABVB, the size must be $720 \times 486$ for NTSC and $720 \times 576$ for PAL. But there is more to it if you are working in a square pixel graphics program like Photoshop. You must work in a square pixel world in Photoshop, which requires an original size of $720 \times 540$ (slightly larger version of $640 \times 480$) for NTSC or $768 \times 576$ for PAL. It is only after all the creative work is done that you squeeze the image by resizing it to $720 \times 486$ (NTSC) or $720 \times 576$ (PAL).

**Figure 6.1** Import Settings

When you import the graphic, the MC/AX will appear to unsqueeze it and it will look normal. With PAL you must still use Force to Fit as an import choice in the Import Settings. The circles will be round and the squares will be square. If you do not do the resizing then the MC/AX does it for you and compensates by stretching the image during the import. It makes the graphic fit, but you will not be as happy with the results.

## Force to Fit and Maintain Aspect Ratio

The 6.*x* and later versions of the MC/AX allow you to Force to Fit, an option that takes any size and tries to stretch or squeeze to make it right. If you have the right aspect ratio it looks OK, but if it is long and thin, for instance, the only way to force it to fit is to squeeze it into the right aspect ratio. This is usually not acceptable and requires some human intervention to make the call on what must be cropped. The resizing of the detail in pre-7.0/2.0 versions may not look as good as if it was done by Photoshop or another third-party program that specializes in image resizing. The difference is subtle, but worth it if there is a lot of fine detail and the finished quality is of the utmost importance.

If you have the wrong size and you do not choose Force to Fit, instead choosing Maintain Aspect Ratio, then the MC/AX creates a black background around the image and makes sure that the largest dimension falls within the correct size. This still may not be acceptable if you have not taken into account the safe title and the safe picture area around the outside of the frame. Everything in the outside 10% of the frame will probably be cut off on most home TV sets. If that information is important, you must compensate for it before importing by pushing important details away from the edge of the graphic. You can also squeeze the image smaller after it is imported by using a picture-in-picture effect.

If you are working with the NuVista+ video board, things are simpler. You just import at 640 × 480 if you are working in NTSC and that is that. You don't have the wider range of video levels or the higher AVRs available to you, but it's hard to screw up. PAL unfortunately has another step with the NuVista+. Just as you must resize with the ABVB, you must resize with the NuVista+ with PAL. Start with the 768 × 576 and then resize to 640 × 576 before importing. Especially in PAL, you will notice a dramatic decrease in import time and increased image quality when you do the resizing outside of the MC/AX application.

So here again are the size rules for importing:

- NTSC: With the ABVB, create the image at 720 × 540 and squeeze the image to 720 × 486 before importing. With the NuVista+, use 640 × 480.
- PAL: With the ABVB, create the image at 768 × 576 and then squeeze the image to 720 × 576 and import with Force to Fit. For the NuVista+, create the image at 768 × 576 and resize to 640 × 576.

## Resolution

Resolution is linked closely to size. If you have an image that is the correct size but the wrong resolution, it will import as if it is the wrong size. Graphic artists may insist that the higher the resolution, the better the image will appear in the final product, but this is not true for television above a certain level of resolution. You have a limited resolution that is always the same: 72 pixels per inch (ppi). It doesn't matter what the original resolution is, it must be converted to 72 ppi. If you have Maintain Aspect Ratio checked and the resolution is wrong, the MC/AX adds a black border around the outside of the image.

## Format

The last characteristic that must be taken into consideration is format, although this is going to be less important by the time this book is published. With existing 6.5.x on the MC and 1.5.x on the Avid Xpress, the only format that can be imported as a still and really work well is the PICT format.

There are other import choices, many more now with versions 7.0/2.0, but some of them are for simplicity and may not give you the results you want for stills. For example, you can import directly from a photo-CD and, if you have no other third-party programs, this is the only choice. But the size of a 35-millimeter scanned slide does not match the aspect ratio of video (square or non-square) and so will either give you a black border if you choose Maintain Aspect Ratio or squeeze the sides if you choose Force to Fit. You really must decide which part of this image to discard (not to mention if it is a vertical composition!); that should not be left up to the computer. Choose what must be discarded from the image and crop.

## Alpha Channels

The other reason to choose PICT at this point is because you can attach an alpha channel to it and use the higher quality anti-aliased edge of this alpha channel to create the best key the MC/AX is capable of. There are other formats that allow alpha channels so with versions 7.0/2.0, there is less of a need to convert to PICT. But make sure, if you need an alpha channel, that the format you use supports it.

An alpha channel should be seriously considered as the method of choice for keying images. It is a fourth channel after the red, green, and blue channels of an RGB graphic. An alpha channel holds grayscale shapes that determine what part of a graphic will be opaque and what part transparent. It is a bit of work, carefully drawing around edges, and could easily be handed off to someone more skilled at drawing or who has a dedicated computer just for graphics. Once you become good at it you may find it easier and faster to just whip one off yourself. Depending on your image, perhaps it is high contrast already, and your skill with the graphics program, you may be able to use some automatic selection functions to create your alpha channel. Using the new matte cutting tools in version 7.0, you can draw a matte in the MC and be able to change that matte over time.

Working with a graphics tablet makes it easier for the fine adjustments to a matte selection made in a graphics program. Make sure the edges you create are always anti-aliased since this always gives you the better keying edge. Some people insist that anti-aliasing should never be added to an alpha channel because it gives a little unwanted border around the outside edge. This is a result of the three-pixel-wide edge that is created in a straight alpha channel when you anti-alias and part of why it gives a nice, smooth edge that ramps from full opacity to full transparency. The three pixels cause the edge to extend one pixel beyond the edge that you have so carefully drawn. Luckily, it is very easy to adjust the edge of the selection and "choke" it back by the necessary one pixel. This is very easy in Photoshop 3.0 and later using Modify Selection, but is tedious in Photoshop 2.5 and is well worth the upgrade from the earlier version. All professional-level graphics programs should allow you to choke the selected edge by a pixel. This clears up any border that is a result of anti-aliased edges of a straight alpha channel.

The standard in television graphics on dedicated graphics workstations, like the Quantel Paintbox, has been to make any mask or hi-con black where the

**Figure 6.2**  The Three Pixels of an Anti-aliased Edge

keyed graphic was to be transparent and white where it was to be opaque. Most video switchers inverted this if necessary, but nonetheless this was common practice in the video world. In the film world, however, opticals were created with the opposite use of black and white and it was this orientation, chosen many years before Media Composer became an online tool, that now prevails. This means that, for simplicity sake, all graphics should be created with an alpha channel that has white for transparent areas and black for opaque areas. Any graphic can have its alpha channel inverted during import or with just a few keystrokes in a graphics program. Confusion arises when some graphics coming from one source are one way and graphics coming from another source are the opposite. This is something that should be specified as part of any facility's production guidelines. The only exception to this would be if these same graphics were going to other, multiple uses and the video use does not merit a second version. Do one test import and see if you have background and no foreground in your keyed graphic. If so, time to use the Invert Alpha Channel during import!

## RGB VS. CCIR FOR IMPORT

We need to refer to the earlier discussion of CCIR (ITU-R) levels versus RGB levels when bringing in our graphic. Chances are, since the graphic originated in a graphics program, it was created with RGB values. This means that the levels of red, green, and blue at 0, 0, and 0 equal the blackest black possible in that program. Conversely, 255, 255, and 255 are the brightest. Now comes the dilemma: How dark and bright should they be in the video program? In the MC/AX, you are dealing with the digital CCIR values. You are mapping the RGB values from the graphics program to two different sets of digital video values. These values are the same for both PAL and NTSC. The first set of values, RGB, imports the 0 levels from the graphics program to digital level 16, broadcast black. The 255 levels will be mapped to digital level 235, which is the brightest broadcastable white. With the CCIR values, 0 from the graphics program is mapped to digital 0 and the 255 graphics levels are mapped to 255 digital video levels. There are legitimate reasons for both, which is why there is a choice during the import process for CCIR values or RGB values.

## RGB Import

First, let's discuss the RGB values. RGB was the default, and the only way, to import graphics up to version 6.5 and is still the only way to import if you are using the NuVista+ video board because it does not have the CCIR abilities of the ABVB. This is actually the simpler, safer way to import because nothing in the video signal will be below the legal levels for black. You are actually losing all of those levels of black if they were in the graphic, but are gaining the confidence that the video levels will be safe for broadcast. It also takes all of the other values and maps them so that nothing is brighter than 100 IRE (or 1 volt in PAL). You would not be able to match a level in the graphics program with a level in the digitized video that went above 100 IRE.

## CCIR Import

Importing using the CCIR option takes the 0, 0, 0 RGB levels of the graphic and maps them to 0 digital. (Technically .74 IRE in NTSC, but who's counting?) The darkest part of the picture is mapped to the darkest possible video level. The highest RGB value, 255, 255, 255, maps to 255 digital, the highest video level. This gets you into trouble during broadcast and the only way to avoid this is to run the graphic first through a third-party, safe-levels filter that makes sure that everything in the graphic was safe both in video level and color saturation. Sometimes these filters can change the hue as a by-product of this change.

But sometimes you don't care if the video levels are safe, you just want them to match something already digitized. Clearly, importing at RGB levels changes your levels and importing the CCIR levels does not, so after you have finished tweaking a shot you may want it to fit seamlessly into the existing video. CCIR would be the way to go and is of absolute importance if you are working with a frame of video exported from the MC/AX that has values below or above the legal ranges. Some would say you should just redigitize the video again and get the levels within the proper specifications, but let's face it, that is not very often a real-world scenario.

Another reason you might not care if your video levels are safe is if you want to use the superblack for luminance keys. To get a really good key, make sure that nothing in the image you want to key is even close to the 0 black of the background that will be keyed out. I have seen people using between 11 IRE and 15 IRE in NTSC on graphics that have subtle shading and don't want the darkest part of their graphic to be keyed out. So if you want something to be imported exactly the way it was exported or you want to make use of parts of the video signal beyond normal broadcast ranges, use the CCIR values choice during import.

## EXPORTING

One of the joys of working on a computer that can contain several professional programs instead of a dedicated workstation is the ability to quickly export a frame from the video application, tweak it in a graphics program, and import it back. No longer do you need to call the graphics department at the last minute to make the minor changes that are inevitable as the deadlines get closer. It makes it easy to take any frame and use it as a background, do some simple rotoscoping (painting on the video frame by frame), or isolate parts of a frame with an alpha channel. The trap, of course, is that you will always be counted on to do this once you have shown how easy it is!

Exporting has its own set of choices, but most of them are decided by the ultimate use of the image. If you are exporting an image to work on it and then you reimport back into the MC/AX, you want the export and import to be exactly the same. This means choosing CCIR video levels and the native frame size. In NTSC that native frame size is $720 \times 486$ and in PAL $720 \times 576$ for the ABVB.

If you change the frame size during export, two things happen. First, it takes much longer to export since the application must do more work resizing. Second, and more seriously, the scan lines are disturbed from their precise, standardized relationship. If you export a still at a small size for use in a document or for a Web page, the scan lines are not as important because the image is not going back to video. A video import back to the MC/AX needs the interlaced scan line information to reproduce the image exactly. If you resize, even to a square pixel size like $720 \times 540$ or $768 \times 576$, then when you bring the image back to a non-square video board like the ABVB, the import process does not know exactly how the original scan lines were laid out. During the import, the software puts the lines in a slightly different order from the original. Forget trying to match back seamlessly to the original video if the scan lines are in slightly different places. You will also experience a loss of resolution as some scan lines are

**Figure 6.3** Export Formats

doubled to make up for missing ones. If you are exporting to reimport, always choose the native frame size and do not resize the image at all in the graphics program.

## DE-INTERLACING

As I mentioned earlier, there is a slight difference between the image in the first scan line and the second scan line a 60th of a second later if there is motion in the frame exported. This is usually unacceptable if the end result is a print or designated for Web use—some kind of a still. One field must be eliminated, leaving only half of the resolution. Since every other scan line must be doubled or interpolated to fill in the empty space left by deleting half the frame, the image appears to be somewhat softer. Photoshop has a de-interlace filter that nicely removes your problem and with version 4.0 and later, you can program this de-interlace as an Action or macro. For Web use this might not be such a big deal since you are going to be reducing the quality of this image anyway in order for it to load across the Internet with better than glacial speeds. But there are enough third-party programs out there that blend scan lines, reduce video noise, and limit the color palette so that you should always be planning on running your graphics through an extra, final step before their end use.

## SAFE COLORS

There is a danger that anything you do to the image in the third-party program may extend the video and chroma levels beyond the legal broadcast range. The first answer is to use only colors and brightness levels that are already in the image through the eyedropper or any other device you have in the graphics program that allows you to sample part of the image and work with that. Remember that digital-safe levels are 16 for black and 235 for white. There is usually a palette in the graphics programs that you can open to allow you read the RGB levels of any individual pixel if you are suspicious.

This technique is not going to help you if you have applied a special effect filter to your image because the filter is not really restricted by safe colors or brightness levels for video. In fact, most people feel the wilder looking the better! This will be a good way to identify the look of the late '90s, but for now it poses the problem of checking each and every image before importing back into the MC/AX. Both Photoshop and DeBabelizer have safe level filters. Many safe filters just lower the brightness level of the image, but since DeBabelizer works with both chroma or brightness levels, it is the better choice.

All of this works pretty well if you stay consistent and keep your expectations realistic. You will never get a great still for print from digitized video that even comes close to a real photograph. Never plan to get your print advertising from freeze frames exported from the MC/AX unless you plan to run serious filters over them for a look that has nothing to do with reproduction accuracy (exactly what I did for the cover of this book).

## IMPORTING AND EXPORTING MOTION VIDEO

The next step for import and export is using moving video. Usually, people want to import animations created by a 3-D program or an effect sequence rendered from a compositing program. There is also the demand to export for Web pages, CD-ROM, or for material to be used in a compositing program like After Effects. All of the already mentioned procedures apply, including resizing and safe colors, but with slight differences. Also you must decide between two choices for formats: QuickTime or PICT sequences.

Let's look at importing video first. The choice of whether to render as QuickTime or PICT sequence depends on several factors. The first and most important is: What format does your third-party application have as an export choice? Usually less complex programs have only one choice and that is QuickTime. This does indeed make things simpler.

## QuickTime

QuickTime has earned a mixed reputation because most people experience it as a postage stamp, pixellated image that drops frames to stay in sync with the audio. This is how QuickTime plays back on most Macintosh computers without help from SCSI accelerators and special video boards. Most professionals shudder when contemplating what their flashy, quick-cut and heavy-effects sequence would look like when reduced to such pathetic playback performance. But you are not going to playback our QuickTime! Not yet anyway. All you want to do with the QuickTime is use it as a transfer medium from one program to another. Since it allows rendering without any further compression, it has become the de facto standard for transferring video in the Macintosh world. Try to play this and you get giant frames that update about every five seconds, but copy it to a drive or send it down the network and it imports clean and shiny, with no loss of quality, into your MC/AX.

A QuickTime movie is a single large file that must stay under the Macintosh operating system file size limit of two gigabytes. Two gigabytes seems pretty large, but if you are working with high-resolution video, for the MC/AX this means between 250 and 300 kilobytes per frame, you will reach that size between around 4.5 minutes and 3.7 minutes. Usually when a project requires a long QuickTime movie for playback, it is a much reduced frame size and resolution. This chapter will discuss that issue later; for now we are concerned with using QuickTime as a high-resolution transfer medium.

### *The Avid Codec*

The key to getting a QuickTime movie rendered at high quality that imports quickly into the MC/AX is to use the Avid codec. A codec is a system, extension that any third-party program can access because it is in the Macintosh system folder along with other extensions. It allows other programs to compress and decompress a rendered QuickTime movie using the Avid media file format. This means that even though you are rendering a QuickTime movie, you are really creating an Avid media file that is inside the QuickTime format. Technically, this is called *encapsulating*. To work best, use the native frame size and do not mix AVRs. Avid is going even one step further to make working with QuickTime easier by working with Apple to make QuickTime compatible with Avid's Open Media Framework Interchange (OMFI). This compatibility allows you to access the individual frames of a media file and work with them in any QuickTime-compatible program. We will discuss OMFI at the end of this chapter.

So make sure that you load the Avid codec into the system folder of any Macintosh you are using for graphics or animation and you will be given Avid AVRs when you choose the quality of your rendered QuickTime movie. Send the codec to your graphics people, or anyone that is subcontracting graphics for you. And by all means, make sure it is the most recent version! You can get that information by calling Avid Support and be aware that the version of the codec may change on a different schedule from the software itself. The reward of using the codec and native frame size is the almost real-time import speed as you come back to the MC/AX.

### *Size*

Size is still a consideration for importing QuickTime and for all the same reasons. Before, with a still image, size was important because of the scan-line relationship and that is still crucial. But there is now another consideration, and that is speed. If you render at anything other than the native frame size, you have to double lines to make up for the difference or throw away resolution and waste rendering time. You are also wasting import time since the MC/AX must resize each frame on the fly and this adds a considerable, unacceptable amount of extra time to the process. If you must eventually create a square pixel frame, consider resizing using another program like Media Cleaner Pro on another Mac (with the Avid codec loaded) so that at least you are not tying up the MC/AX while the QuickTime movie is resizing.

The other choices involved, when rendering to import to the MC/AX, have to do with field order and field rendering. To get the absolute best quality rendered movie, choose field rendering if your application makes it available. When you make this choice you must also choose field one to have priority, or be rendered first. Field rendering takes longer to render, but is worth it if you are working with complicated video and lots of detail and you want this to be a finished product.

### *Importing with an Alpha Channel*

One option for QuickTime that is not yet supported by the Avid codec is alpha channels. Although alpha channels are a simple matter with still images, they are much more complex with moving alpha channels and moving video. You are, in effect, importing two separate streams of video at the same time. You have two choices if you want to import an animation with a moving alpha channel.

The first choice is to not use the Avid codec, but instead to choose two other compression formats that support alpha channels: animation and none. When you choose which type of compression to render your animation or composition, there are many, confusing choices, but you should use these two for finished quality work if you are not going to use the Avid codec. Animation is designed for what it is named, animation, and uses a lossless compression if you adjust the quality slider to its highest level. Because it is compressed it saves a little bit of disk space and you will not hit the two-gigabyte file size limit quite so quickly.

The other choice, none, means that there is no compression at all. This results in larger file sizes and longer transfer or copy times, but there is nothing between you and your video.

When you import these QuickTime movies with alpha channels into the MC/AX, you end up with a two-layer matte key effect that must be rendered. This requires a relatively long render time in the MC/AX because the matte key is not a real-time effect and cannot take advantage of the fast render capabilities.

The other choice for QuickTime animation with an alpha channel is to use the Avid codec, but render the animation as two separate files. Render one pass for the animation and another for the alpha channel. If you can set them up to render one after the other and you are going home for the night anyway, the extra rendering time may make no difference to your schedule. However, you will still pay a time penalty when it comes time to render in the MC/AX. You must take these two separate imported animations and layer them on top of each other, putting the alpha channel on top, and then apply a matte key effect to the alpha channel. It takes only a few seconds to assemble this effect, but the rendering time for a matte key is considerable.

## PICT Sequences

Most people choose QuickTime as their medium of choice for animation because it is just one large file that is easy to keep track of and because it is the early universal choice of most animation programs. There are times when the animation program may originate on a computer platform that doesn't support QuickTime, such as a Sun Workstation, or you are choosing to work in a graphics program to affect each individual, exported video frame. In these cases you want to work with a PICT sequence. If you are using Avid's Illusion, however, you can output as OMFI and simplify the process.

There is a little twist involved in getting PICT sequences to import to the MC/AX and it has to do with file naming. Any PICT sequence must be recognized by the MC/AX as a sequence and not as a series of unrelated PICTs. The way to be sure that you have a PICT sequence is with the file name, which must follow a strict format of: Name.001, Name.002, Name.003, etc. The name must be followed by a period and a number. Before versions 7.0/2.0, "Name.001.pic" was an invalid name and the "pic" needed to be removed. With versions 7.0/2.0, the extra part of the name after the number is accepted correctly, but it must be removed before import for earlier versions. With the pre-7.0/2.0 versions, you must get the PICT file name right or you will be faced with running every file through a renaming program like Knoll Renamer or creating a macro in DeBabelizer. These are not extremely time consuming, just tedious and distracting from the work at hand (especially if you must track down one of these renaming programs while the clock is ticking).

The MC/AX will look for ".001" to be the first image in the animation. If you have started with ".000," versions 7.0/2.0 will give you an exception message. If there is no ".001" it does not see the rest of the images as a sequence. The only way to get the MC/AX to see numbered PICT files as a PICT sequence is to check the Import sequential files check box in the Import Setting. If each image comes in as a 10 second master clip, you know you forgot something! The system has imported each of the PICT sequence frames as a separate, conventional PICT. The finished sequence should come in as one frame per image with 30 frames per second NTSC and 25 frames per second PAL.

The main benefit of working with PICT files to import into the MC/AX is that a PICT file supports an alpha channel and can be imported that way. Importing automatically sets up a matte key effect although, once again, because it will be moving, the matte key effect is not real-time and must be rendered. However, it is simple and relatively problem free. Remember that the alpha must be inverted from what most programs create as a default. Remind your animator or invert on import. Either way make sure you know whether you have white on black (need to invert) or black on white (import without adjustment) before you import so you don't do it twice!

Consider how easy it is to set up a network between an SGI and your MC/AX as an Ethernet or faster TCP/IP. TCP/IP is a protocol that you can use between two computers in your facility (technically, an intranet if done this way). This simplifies the transfer between computer platforms admirably and you no longer need to worry about a removable drive, compatible hard drive formats, or the time to copy and restore. Besides, most SGI workstations don't use a floppy drive!

The original source for your PICT sequence may have come from the MC/AX and was exported as a PICT sequence. In this case, you are dealing with many individual PICTs and may be using these to go to an SGI or Windows NT-based program that doesn't use QuickTime. Or you may want to paint or draw on each individual frame and paint out wires, boom microphones, or ex-spouses. This is time-consuming because of the huge amount of work involved in 30 or 25 hand-painted frames per second. It is the ultimate "fix it in post" and last resort for fixing problems created in the field, but the results can be magical if done right. This is exactly the kind of work the intraframe painting tools of version 7.0 were designed for.

A simpler way to affect each individual frame is to apply a filter to each one in a macro-based program like DeBabelizer. Set up the macro to open each frame, run an effect filter, run a safe colors filter, and then save the new frame with a different name. Of course, After Effects can do this to a QuickTime movie too, but there will be rendering time. Now that the new Avid format, called AVX (Avid Video eXchange), is available, this will simplify the filter process even more. You will be able to buy a version of your favorite filter that has been converted to AVX and apply it through the MC/AX like any other effect. Early Photoshop plug-ins for version 2.0 actually show up in the Effects palette when you drag them into the 3rd Party Plug-Ins folder inside the Avid Supporting Files folder. Photoshop plug-ins took so long to render, even a single frame, that they were generally impractical.

When creating graphics from scratch to be imported into the MC/AX, remember that they must be resized to the non-square pixel format if you are working in a square pixel graphics program. Hopefully, the animation program you are using works with non-square pixels, but if not, you must build every frame in the square pixel size and resize before saving. If you have been handed a square pixel animation by a subcontractor, you should resize it yourself. Again, a macro-based program like DeBabelizer or Media Cleaner Pro will come to your rescue. Alternatively, you can place the square pixel animation in a D1 size composition in After Effects, resize there, and render (don't underestimate the rendering time).

## Media Cleaner Pro and CD-ROMs

The MC/AX allows you much flexibility with exporting QuickTime, but depending on your needs, you may still want to use a third-party program like Media Cleaner Pro by Terran Interactive to do the final compression for playback. Cinepak is the type of compression most multimedia software needs. From the MC/AX, you can export at different frame rates, frame size, and compression choices so that you could play back QuickTime straight from the Macintosh desktop or go straight to CD-ROM. Except for short movies, though, you may want to export everything using the Avid codec at full frame rate and native frame size. Send the QuickTime file to another Mac workstation where Media Cleaner Pro can churn away in batches and create the best Cinepak-compressed version with blended frames and compressed audio.

The exact procedures for creating the best CD-ROM from an MC/AX have been written and published before by Avid and written by Greg Staten. You can find it on the Avid Web site (www.avid.com) as a past Avid Answers article. It is too detailed to go into here, but if your job is to get the best CD-ROM playback from your system, it is well worth reading. One thing to keep in mind is the extremely limited ability of a CD-ROM to playback QuickTime—it is a very slow drive. You have a maximum rate of between 175 and 200 kilobytes per second for a 2X CD-ROM player. Yes, there are 24X players available (which will be outdated by the publication of this book), but you can not count on these for a wide distribution of a commercial CD-ROM. If you are distributing to a controlled, in-house group, then you may go for a higher playback rate. Some people even have the luxury of loading their video or multimedia presentations onto a central server and are limited only by the speed of the hard drive and the network itself. Media Cleaner Pro allows you to make a series of intelligent decisions and weigh the pros and cons of your quality choices and the playback performance.

### *Cinepak*

The compression level is probably the most important decision besides the frame rate for determining the data rate of a QuickTime movie that has been compressed for CD-ROM. By far and away the most popular form of compression for playback on a computer without any special hardware is Cinepak. Cinepak is useful because it not only compresses each frame like motion JPEG, called *spatial compression*, but compresses over time as well, called *temporal compression*. Temporal compression requires reference frames, keyframes that are not compressed over time. The more keyframes you have the better your video will look. Because the compression has determined that certain parts of the frame do not need to be updated until the next keyframe, if there are sudden, unexpected changes like a series of quick cuts or abrupt action, Cinepak creates natural keyframes when the majority of pixels in an image change. So why wouldn't you have huge numbers of keyframes, for instance, every other frame? Because that defeats the purpose of compression over time. The keyframes are a reference, take up more space, and require a higher throughput. It is because there are all of these smaller, compressed frames that you save space with Cinepak.

### Audio for QuickTime

Audio is also a very important consideration for finished QuickTime playback. If you are exporting to a compositing program like After Effects, you are better off using the audio just as a reference so that you know you are hitting the right moments while creating your effects. But low-quality audio is not acceptable when the final QuickTime is pressed to CD-ROM. The problem again is with data rate. What will most users respond to: jerky and blurry video or compressed, mono audio? The answer most of the time will be that the quality of the video is more important than the quality of the audio. Unless the sound is distorted or unintelligible, people are happy playing lower quality audio through the little speakers in their computer. Media Cleaner Pro offers a good compromise for compression called IMA (Interactive Multimedia Association) that can be played back from all but the oldest Macs.

There are a few issues that have to do with volume level with exporting audio from the MC/AX. Again, evaluate whether your audio is important enough to double its data rate when the majority of listeners will be listening to it in mono anyway. Mono export means that you need to make a duplicate of your sequence and prepare the audio for mono playback. Pan the audio either all to the left channel or all to the right channel and raise the audio levels as high as possible without distorting. Then make an audio mixdown. Delete the original audio tracks, leave the mixdown track and export.

The future of importing and exporting is now much simpler for the MC/AX. The choices for import will be for shot log, graphics/audio, OMFI, and EditCam. With graphics, the MC/AX takes a look at your graphic and determines what format it is and whether it can import that format. This removes a lot of the complications involved with having to change formats. This new ability is because Avid has incorporated the technology called HIIP (Host Independent Image Protocol). There is auto-detection of file type using HIIP in versions 7.0/2.0 and there are now over 25 formats that you can import and export.

## OMFI

A cross-platform choice for importing and exporting that is now used extensively is OMFI (Open Media Framework Interchange). It is a format that Avid developed but is non-proprietary and over 400 manufacturers and software companies are using it. The beauty of OMFI is that it can contain important information about your Avid sequence. It can contain both media and composition information and can represent a much greater complexity than an EDL. It is the way that other systems input all of that complicated layering, effects, audio mixing, and links to digitized media. It keeps you from having to reproduce it all. If you are going to be exporting sequence information with your media, this is the way to go and any facility that is looking to exchange large amounts of media across platforms and across applications should seriously look into using OMFI. The standard edit decision list is almost useless for representing complicated effects information across different manufacturers of hardware and software.

### Using OMFI for Pro Tools

OMFI is used extensively for moving audio from a video session on the MC/AX to a digital audio workstation. If you are moving the audio to a Pro Tools session there are two methods to consider. The first method requires less drive space and is faster, but the second method allows more flexibility if you are going to software that does not recognize the Sound Designer II format that Pro Tools and MC/AX uses. AudioVision supports a direct translation from the Pro Tools session so once you convert it for Pro Tools, AudioVision can read it too. Versions 7.0/2.0 allow an OMFI export as an AudioVision compatible file to skip this middle step.

The first method is to hand over the drive with the audio media files on it and create an OMFI file that is composition only and also an SD 2 format. The OMFI holds all the edit information and the drive containing all the audio files must be moved from the ATTO SCSI chain to the Macintosh SCSI chain. You may want to consolidate your sequence before you do this if you want to put all the audio media on another drive for transport to the audio workstation. This allows you to keep working with the audio files you have, although you might want to lock your audio tracks in the sequence so you don't accidentally change something while the mix is going on.

The second method is to make an OMFI Audio Only file. This creates an intermediate file that is actually another, more widely used audio format, AIFF, along with all the edit information. Once this very large file is moved over to the digital audio workstation, it can be converted back to an SD II file for use with the Pro Tools software. The intermediate OMFI file is opened up in the OMF Tool that comes with Pro Tools and converted to a Pro Tools session. This OMFI file can be deleted after the conversion.

In order to send this audio mix back to the MC/AX you must "bounce" the audio tracks to a continuous audio track. This is a real-time process that changes the audio file, which now has subframe edits, to a frame rate that can be used by the MC/AX. Alternately you can output to a digital tape format and redigitize it into the MC/AX and line it up to the beginning of the sequence. Having some sort of synchronizing beep tones with a countdown or flashframe simplifies this final sync.

## ADOBE AFTER EFFECTS

When preparing a composition in After Effects there are certain settings you should always use when rendering for the highest quality. The D1 frame size is the right size for import to the ABVB. If you are bringing frames into After Effects from a square pixel graphics program you should import them and then "auto fit" so that it is resized correctly. Use the Avid codec and use field rendering with upper field first at the speed of 29.97 for NTSC and 25 for PAL. It's up to you to decide whether you want to render with an alpha channel or two separate QuickTime movies, the alpha channel and the video.

Importing and exporting video, audio, and graphics has many variations, formats, and choices. With this flexibility comes complexity so any production company should find the processes that work best for them, simplify them as much as possible, and be aware of that you have many tools at your disposal.

# 7

# Preparing for Linear Online

Most of the preparation for a linear tape-based online happens at the end of a non-linear project when you make the edit decision list (EDL), but there are things that must be done properly throughout the job in order for the online to go smoothly. The reality is that with all the twisted financing that goes on in this crazy industry, there are still reasons to buy a bunch of offline models and create an EDL when finished. For those who have bought the lower end models, which do not have the finishing quality AVRs, you need to prepare for a linear online edit session at the end of every project.

This chapter will discuss the most basic requirements to get a good EDL under the most common scenarios. Then we'll go a bit deeper into the possibilities for increasing the speed of an online auto-assembly, the Holy Grail of all expensive online sessions, and almost as tough to find! But the good news is that many facilities want the list in a simple, bulletproof form so that they can do their own, more complicated variations.

How do you transfer a 24-layer effects sequence with nesting, color effects, audio equalization, and rubberbanding to a format that supports one layer of video, four tracks of audio, and hasn't changed much in almost 20 years? Answer: not very well. Dedicated hardware is the rule in traditional online suites with a different piece of equipment and, many times, a different manufacturer for each function like text, special effects, and tape control. Many times the separate pieces of equipment were purchased years apart and there may be components that are ten years old that still work fine for what they were designed to do. Compare the processor of a CMX 3600 with a Macintosh 9600. The Macintosh 9600 has been engineered so that you can replace the entire processor by changing a single card. Yes, if you design the hardware to do just one thing then you can maximize it so that it doesn't need the RAM or processor speed of a system required to do many things, but just try to add new features! This is why there are so many third-party EDL management software choices. If the dedicated linear online systems cannot add new software features, then new third-party software does it and simplifies the EDL to a format the dedicated systems can handle.

Even with third-party EDL programs there is still a massive mismatch in capabilities between what can be done by clever folks inside the MC/AX and what can be represented by this limited format for a linear online. The only way to convert from one format to the other, preparing for that precious linear online time, is to dumb everything down. You need to understand the limitations of what can be conveyed in an EDL and bring the Avid EDL to a point where the show can be built back up again, piece by piece, in the order necessary in the linear, tape-based world.

This is more of a challenge than it would appear at first, not because it is that difficult, but because of the limited experience of the Avid editor in the world of linear online. Unless you have actually worked as an editor both at an Avid system and at that particular linear online suite, you are not really in a good position to take advantage of all of that tape suite's advantages. We can discuss capabilities in generic terms, but that doesn't help you if you didn't know that, for that suite, you needed to reserve the character generator a day in advance and you can forget using the digital effects unless you book a night session.

This is where you must rely on two very important low-tech tools: clear communication and the ability to collaborate with experts. As you devolve the Avid list to something simple for the linear assembly, you must be very clear exactly what must be done to every part of your sequence. What effect do you really want? How are you going to deal with all those channels of audio? This chapter will describe different ways to leave a paper trail that someone else can follow. Count on the expertise of the operations or scheduling people at the particular post-production house that has been chosen. Speak to the editors if they are available. Depending on time available and the complexity of the program, you may even want to (gasp) plan a pre-post-production meeting and discuss approaches with the editor! Imagine, warning them what to prepare for! Let's face it, there is no way to predict everything that can happen after you leave a project. It is really not fair for the offline editor to be blamed for changes that occurred after the cut is locked, changes that ripple backwards to a decision you made in good faith with the information available weeks or months ago. Unfortunately, this happens all the time.

Coming from a post-production facility background, but having freelanced for a few years as well, I strongly advocate building a good relationship with a handful of production companies. Bring any work you can to these select production companies, large projects and small, and spend some time learning the facility's capabilities and quirks. If you create a sense of loyalty, they will look out for you. This happens in little ways, like working a little harder to fit you in for emergencies or coming to know your work and style and able to anticipate and fix problems for you before you even know they are there. If they can get you in and out on time and on budget, they can fit in more paying sessions, need fewer "make goods" for questionable mistakes and lower their already unnecessarily high stress level.

If you can convince your employers to pay for you to sit in on a few online sessions, it will pay off for them in the long run. If they have a relationship with you, then it is in their best interest for you to continue to make informed decisions about effects that can be recreated easily or faster ways to assemble the show in the final stage. Will it be faster to dub shots onto a selects reel? Is it better to have more tape decks available for a shorter time or fewer decks for a longer time? And really, what can you do at a lower rate per hour to make things go smoother? When you learn these things, you become more valuable to your employer, have more job security or stay on the "A List" the next time your favorite client, director, or producer has a big job.

So it is in everyone's best interest that you understand the implications of every step of your offline on the final online. If you digitize without timecode or without paying attention to tape names, you are directly hindering the next step. Don't blame "those Avid EDLs" when you have no regard (and, really, no respect) for the next link in the chain. It is your job to continue to inform the director or producer that choices they are forcing you to make will cost them money later. If they take just a little more time now, at the lower rates, things will go smoother at the crunch time.

What are the biggest mistakes? Ignoring the importance of timecode and tape name. Everything must have a timecode reference, all sound effects and music, all graphics and animations, and all picture and sync sound. Let me say this again: If you are going to a linear online, every source must have timecode. It is tempting to just import that one cut from the CD music library or just pop in those three shots from the VHS dub, but you must be sure to go back and match by eye or ear with a timecoded original source before you make the EDL. This is a clear example of GIGO—garbage in (sources without timecode) and garbage out (a list which can not really be assembled).

## PREPARING SOUND

There is one exception to the timecode rule: when the audio digitized into the MC/AX will actually be used as a source in the online (or the final mix for a film). It is becoming more common for the first edit to lay down a digital cut of the audio, low-resolution video and then cut with original video sources the rest of the day. This assumes no changes in the online that will affect sync or length of course, something you may have to mention repeatedly! In this case, you are working from the timecode of the digital cut, which should match the timecode of the master sequence. If the show starts at 1:00:00:00, then the sound source should too. Editing the high-quality digital sound works especially well with timecode-controlled DAT or another digital tape format. You could go out of the Avid using the digital audio output format to a DigiBeta tape. In a pinch, audio tracks 3 and 4 of a Betacam SP can be used since these audio tracks have a higher dynamic range and lower signal-to-noise ratio than tracks 1 and 2. This is good for effects and sound under but should not be the main method for the most important sound in the project. Just make sure the Betacam SP decks in the online suite are models that can play tracks 3 and 4. Generally, this means a Sony editing deck with the letters BVW in front of the model number. The non-standardization of high-end Betacam SP decks makes this problem common in lower cost suites. Did I mention you can't make changes?

## USING THE OFFLINE CUT IN ONLINE

If you are using digital videotape to bring the Avid digital cut to the online suite for your audio, then you should also include the video as part of that first edit. There is a clever trick that many online editors use where they lay down both video and audio from the Avid to the master tape in the online suite and put a big circle wipe with color bars in the center of the video. That way they can cut over the top of the old video all day and see if there is any discrepancy between the EDL and the digital cut. Anytime they see a flash of color bars they know something is a frame or a field off and can make adjustments. It sure is better than squinting all day to tell the difference between one frame of AVR 6 and the original!

The other way that the digitized audio without timecode from the MC/AX can be used is when you make an OMFI export of the audio and the sequence for use on a digital audio workstation. But if the original sound was digitized with a little distortion or mono instead of stereo, then the sound mixers have to redigitize it on their workstation and match it back by ear. Not too much trouble if it is just a few cuts, but if the music comes from a five-minute cut and is used at every transition, then you are costing your client money. How do you think the sound guys will justify their higher bill?

When you are dubbing everything to timecoded sources, it is up to you (or your assistants, dubbers, etc., but really up to you) to make sure the quality is not degraded. Dub to higher formats, digital formats, and monitor, the levels very closely. There is nothing worse than being forced to redigitize a distorted media file and finding that the dubbed source is also distorted! Some facilities make sure that everyone who dubs a source put their initials, date, and machine used on every tape. Being able to track a problem back to the source does a lot to ensure quality control.

## DUBBING WITH TIMECODE

There are some basic requirements for good timecode that are very important when dubbing sources to be used in an online session. The first requirement is:

Never dub timecode. That seems like an outrageously stupid assertion, but one that has confused people for many years. You should not dub timecode—you should always regenerate it. Timecode is a square wave signal and straight dubbing tends to add noise to the signal. Eventually (sooner if the timecode wasn't great to start with) the signal becomes slightly rounded and suddenly, where that sharp edge in the signal delineated a specific number, it is now just a little too rounded to read accurately. You may have been dubbing timecode for years and sending your problems downstream or, luckily and more probably, your record deck has been left in the position for "regen." This is a switch under the front control panel of the deck that allows you to regenerate, "re-square," the signal as it passes through the deck's electronics and lays it to tape.

It may be important to keep the timecode from the original source for accounting reasons, the stock footage must be accurately paid for, or because of the way the selects reel will be made. When making a selects reel, you can be dubbing many shots from separate reels onto one reel for the convenience of digitizing or because you will lose access to the originals during post-production. The shots on the selects reel may need to be traced back one more step to the original tapes before the online. Now you should set the internal/external timecode switch on the tape deck to "ext" for external source. Make sure that the source deck and the record deck have timecode cables connected between them so that you can jam sync the timecode to the new tape. Give yourself plenty of preroll because there is a break in the timecode at every shot and you do not want the Avid or the edit controller to rewind over a break in the timecode when cueing up for a shot.

All edit controllers, whether they are part of the Avid software or part of an online suite, always expect a higher number to come later on the tape. The timecode is expected to increment continuously! If you put hour 10 before hour 5 on the tape, anyone who uses this tape will be confused. Edit controllers will search and fail to find preroll and edit points. Your name will be used in vain and no one will ever return your phone calls again.

If the beginning shots of the tape are drop-frame then everything that is dubbed to that tape should be drop-frame too. The MC/AX associates a tape name with whether an entire tape is drop or non-drop. If you mix the types of timecode you have to give the tape two names—very confusing and not recommended!

If you can lay down new timecode instead of taking it from the original source, you can keep the timecode nice and neat on the selects reel. If the "ext/int" switch is back in the "int" (internal) position, the record deck always generates its own timecode. The beauty of also leaving your deck in "regen" is that any assemble edit (an edit that changes the timecode on the tape) picks up where the last timecode on the tape left off. During the few seconds preroll to make the assemble edit, the deck is reading the timecode already on the tape and neatly regenerates it at the edit point without missing a frame. Either you should be purposefully presetting the timecode on the deck, jam synching the code from the source tape, or regenerating it from the timecode that is already on the new tape.

Timecode is just another tool and can be used very effectively once you know how to preset it and actively use it. If you are making a selects reel, choose a timecode that is not used by the other field tapes. If the last field tape is tape 15 and uses timecode hour 15, then start your selects reel at hour 16. No matter how a tape is named in the EDL, you can trace it back to the right tape based on the hour of the timecode.

## TAPE NAMES IN EDLS

Tape name is incredibly important to the MC/AX and, if you get almost nothing else right, at least this should be perfect. Every tape needs a unique name and must be named while digitizing or logging in the correct project. It is good practice to make the tape name match the timecode hour and never repeat a tape name. As mentioned before, but bears repeating, tape names should be short— five to seven characters maximum to allow a space for a "B" to be attached to the end when making dupe reels for CMX and Grass Valley Group (GVG) systems. This chapter will discuss dupe reels later—you can't count on not needing them!

Assuming that all timecodes in the MC/AX refer back to real sources and all tape names are unique, let's look at the basics for preparing a list. The Media Composer has been shipping with an extra piece of software for the last few years called EDL Manager. It also ships with the Avid Xpress; if you don't have it on older versions check the box your User Guide came in and you will probably find it there neatly shrink-wrapped. This piece of software is now part of the installation CD-ROM and is meant to be able to open bins and locate the sequences inside the bins to create EDLs.

## EDL MANAGER

EDL Manager is Avid's EDL generation and list-importing software and is separate from the MC/AX software for several reasons. The first reason is so that it can be updated and improved upon with a development schedule separate from the Media Composer. EDL Manager can then skip the long and complicated beta testing programs in favor of quality assurance just for that small piece of software. The offline editor can carry it to the online suite or the online editor can have a version running in the suite already. The offline editor can be spared some of the responsibility for making a good list by bringing a sequence bin to the online session with the (hopefully) final version of the sequence.

With present versions of the MC/AX, the EDL Manager launches automatically when you choose EDL from the Output menu. If it doesn't launch, there are probably two things wrong. First, the EDL Manager application must stay at the level of the Macintosh file hierarchy where it was originally installed. In other words, it cannot be inside another folder hidden away somewhere. It must be on the internal Macintosh drive on the highest level, visible after you double-click on the drive icon, and inside a folder named "EDL Manager *f*." Second, if it is there and still is not opening, you may not have enough RAM for it. EDL Manager operates now with a minimum of 6 megabytes of RAM, but if you are working on very long sequences you have to give it more, much more. Some people give EDL Manager 40 megabytes of RAM and then never have a problem with any sequence, but they must have the extra RAM available if they are simultaneously running MC/AX.

If both applications are running, you can use the simple arrow icon in EDL Manager to load the sequence in the record window of the MC/AX to the EDL Manager with a single click. You can also use the arrow to go the other way and turn an EDL into a sequence, but since there is more art than science to that technique, we will discuss it in Chapter 8.

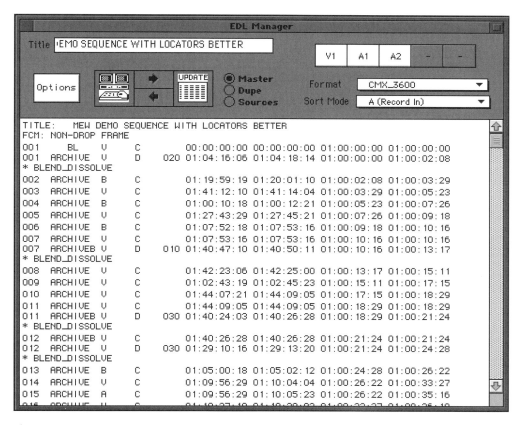

**Figure 7.1** EDL Manager

If you are using earlier versions of the Media Composer, the EDL Tool is built into the Media Composer application. With EDL Tool you are missing several important Sony list types, a standard way to get a well-tested, preread function and a superior EDL import ability. We will talk about the importance of all of these later in this chapter, but with EDL Manager, you are working on a version of software that has been updated and beta tested for years longer than the EDL Tool. If you have a choice, use the EDL Manager.

If you are using the MCX for NT, you have the MCX EDL to use for the same purpose. This application is actually a modified version of a well-known, popular program called EDLMax written originally by Brooks Harris. Harris modified it to make it more compatible with the MCX -NT and OMFI. This discussion concentrates on features that are important to both pieces of software and to any EDL, although the actual menus are different with MCX EDL.

## GETTING READY

There are so many variables in the linear online suite that you can never be too careful about preparing for any eventuality. There are many different kinds of edit controllers on the market today and to some extent they are not compatible. The main ones in use are CMX, Grass Valley Group (GVG), and Sony. There are also many other less prevalent models like Ampex ACE and Abekas, although within your market there may be an exception to this and you may find that the most popular edit controller in your market is something else. In Chicago, for instance you will find lots of Axial controllers.

The first rule of EDLs is this: Find out which edit controller format the linear online facility needs. Notice I said, "what they need" not "what edit controller do they have" and this is for a very important reason. They may have a CMX Omni not a CMX 3600, but the lists are mostly compatible so the facility may ask for a CMX 3600 list instead. Or they may have a bizarre 10-year-old Brand X model that went out of business 7 years ago, but the darn machine just refuses to die. They may ask for a CMX 3600 as well because out of necessity they may have or may have written their own software to translate from CMX 3600. Most people consider the CMX 3600 the Latin root of all list formats.

## FORMATTING FLOPPIES AND RT-11

You must format the floppy disks for CMX and Grass Valley Group edit controllers in an unusual way. CMX originally used a disk format that was useful with the DEC computers they used internally when they were developed. Because GVG developers came from CMX, Grass Valley Group also uses a version of that original CMX format. The format is called RT-11 and is problematic with both Macintoshes and Wintel because these operating systems are not designed to recognize it. Special commands must be written to make the computer ignore the fact that it can't read the RT-11 floppy disk you just popped in. The edit list software must take over for both reading and formatting these floppies. This is why if you just pop a CMX disk into the Macintosh, you get a scary message saying this disk is unrecognizable and do you want to format it? No! Eject, now! The Mac defaults to this evil question whenever it cannot recognize the formatting on a disk. You can only work with RT-11 disks when you are in the MC/AX, EDL Manager, or another piece of specialized software designed specifically for EDLs.

So RT-11 was not written to make your life a living hell, but it sure doesn't help. Add to this the restriction that the disk used must be a double-density, double-sided disk. I want to say that again: double-density, double-sided. Look at all of the floppies strewn around your computer. I'll bet none of them used for standard every day saving is double-density, double-sided. They probably all say HD, which stands for high-density. The disks you want should say DD on them. They are not interchangeable! This is an example of dedicated hardware being passed by very quickly by technological standards. Double-density disks only hold 800 KB of material. This could be a small novel, but it is not very many EDLs. The real world uses HD floppies because they hold 1.4 MB, almost twice the capacity of the DD disks. Even the HD disks are pretty pathetic in a world where gigabytes are becoming the standard measurement for storage capacity. If you are making the edit list, it is up to you to scrutinize every disk that is handed to you.

I strongly recommend the DD disk should be unformatted or already formatted for RT-11 from the edit controller that will use it. There were some early versions of EDL Manager that did not let you format RT-11 if the disk was formatted for anything else already and that was changed for version 1.9.1v2 and later. No matter which version of EDL Manager you may have, there is a trick for unformatting a floppy. First, quit all the programs you are running. Pull down the Erase Disk command under the Macintosh Special menu, as you would normally erase all the material on the floppy. Let the process run for a few seconds, but before it finishes you must force the Macintosh to crash. The only way you may be able to interrupt the erase is to press the restart button on the outside of the computer case or press the Control-Command-power keys all at the same time. This works to totally screw up the format on the floppy, but it sure is ugly.

The formatting of the disk for the EDL has caused many people the vast majority of their EDL woes. I highly recommend getting a floppy that is formatted by the online system that will use it. Barring that, I do not recommend formatting an RT-11 disk on the Avid unless it is an unformatted disk that has never been used for anything else before.

So now you have an unformatted double-density, double-sided floppy. If you pop it back into the Macintosh, the system tells you the floppy is now unrecognizable—this is one of the few times you want to see this message! Now the floppy can be reformatted while you are running the EDL Manager or other EDL software. Is it any wonder that busy post-production facilities would rather FedEx you a few of their own disks? Overnight mail with a proper set of disks is a good idea from their point of view so give them the chance by letting them know you are coming far enough in advance. Sometimes there are differences that have developed over time between the disk drives at their facility and your brand-spanking new Macintosh 9600 floppy drive. Old drives get cranky and may occasionally not accept a floppy that has been formatted by another drive. The busy facility has also seen more than their share of HD disks with EDLs and offline editors that blame the Avid! If all else fails, to get your floppy to be accepted by their online edit controller, copy it to another computer that can read RT-11 (potentially with EDL Manager running), turn the EDL into a text file using a word processing program, and copy it to a properly formatted disk.

## SONY AND DOS

One way to avoid all of this is to work with a Sony list. Sony saves to a DOS formatted disk that the Macintosh can easily handle. Early versions of the Macintosh needed one of two different extensions, either PC Exchange or DOS Mounter. These extensions need to be enabled in the Extensions folder to read or format a floppy for Wintel. You would choose Other during the formatting dialog on the Macintosh and choose the proper format for the size floppy you have. Early Sony edit controllers may only read DD disks, but the later ones read HD. If in doubt, get the DD and format 800K.

## REAL-LIFE VARIATIONS

For one final twist, there may be weird combinations of formats for disks and formats for edit controllers. The later CMX controllers, like the OMNI can accept DOS-formatted disks. The production company may have their own list optimizing software that runs on a DOS or Windows system so they may want the list, no matter what it is, on a DOS floppy. Some older edit controllers still work from 8-inch floppies. Most people have never even seen these before. Forget about compatibility. These unlucky people have devised a way to send information serially, through a cable from another computer that can handle the 3.5-inch floppies to their online system that can only handle 8-inch floppies. EDL Manager allows this too, it just takes a 1-2-3 Go! procedure that you probably don't want anyone to see.

## BE PREPARED

So there appear to be many ways to get this whole EDL process wrong because there are so many ways to get it right. With all of this potential for 9 AM thrash at the online facility, there are ways to cover yourself. First and foremost, you should ask the production facility what they want. Asking far enough ahead of time gives the facility time to send you their pre-formatted floppies just in case. If you are using the same production houses over and over, then you know their drill and can have procedures in place for each edit suite of each production company you use.

But don't bet everything on one floppy! Make two, they're cheap! You may also want to save the EDL in several other formats. Always make exactly what the facility asks for and then make a CMX 3600 list. Pick your market's favorite format and you should make that your backup format. It is up to you whether you want to put it on DOS or RT-11. (If you choose both, it could make you look pretty slick.) Then you should save the sequence bin to a Macintosh-formatted HD floppy and, if the project was small enough, include all the bins. You may find yourself actually making the list again from the sequence at the online facility's Avid or from your own copy of EDL Manager (which you brought with you). After all of this is done: Trust nothing! Make two printouts and bring a copy of the project on VHS or some other format that can be played back in the online suite without a charge for another deck. You or your producer should control the playback of this dub in the suite so you can follow along with the online assembly.

## WHAT IS AN EDL?

Now that you are prepared to make the EDL, let's take a look at what an EDL really is and why it is so hard to make one from a nonlinear editing system. The only information the edit controller wants from the Avid is the tape name and timecode (sound familiar?). The edit controller cues up the source tape and inserts the shot onto a specific timecode on the master tape. So a basic event, a cut for instance, needs 4 timecode numbers: an inpoint and an outpoint on the source tape and corresponding in and outs on the master. It looks like this:

```
017   100      V    C          02:00:24:19  02:00:25:14  01:00:22:02  01:00:22:27
```

When you need to make a dissolve with analog tape, most of the time you need a second source. If you must dub a shot onto a second tape it is typically called a B Roll. Thus, an A/B roll system is one that allows dissolves. A dissolve needs the inpoints and outpoints from the B Roll in the event and looks like this:

```
012   100      V    C          02:18:31:29  02:18:31:29  01:00:12:19  01:00:12:19
012   100B     V    D    030   02:23:26:15  02:23:28:26  01:00:12:19  01:00:15:00
* BLEND_DISSOLVE
```

Any kind of fade up from black or from a graphic that has no timecode is an event that starts with black (BLK) or an auxiliary source (AX or AUX). It looks like this:

```
002     AX      V       C        00:00:03:27 00:00:03:27 01:00:01:28 01:00:01:28
002     100     V       D    030 01:43:44:28 01:43:50:29 01:00:01:28 01:00:07:29
* BLEND_DISSOLVE
```

So all of your edits that come from a non-timecoded source like CD audio or a PICT file come in as AUX. If you have 100 PICT files this could be a problem.

## TRANSLATING EFFECTS

Any time you create an effect like a wipe or a 3-D warp, it must somehow be translated to this very limited, older format that is concerned with tapes and timecodes. There are standard SMPTE (Society of Motion Pictures and Television Engineers) wipe codes, but some of the most popular effects switchers do not use them. EDL Manager has a list of the most popular switchers represented and you can use these simple templates to represent your wipe effects. You can even take these templates for the switchers and modify them to your own switcher if you are feeling geeky. The wipe codes can usually be found in a manufacturer's technical manuals and these templates are just text files.

There is no way to create a generic 3-D effect that any group of DVEs (digital video effects generators) made by different manufacturers can read. Many people don't have the capability to send DVE information from their edit controller to the DVE. These two machines could easily be from different manufacturers and have no communication between them except "Go Now." Different DVEs use different scales for computing 3-D space, their page turns look incredibly different (if they can even do a page turn), and really, forget it, there is no standard for effects and there will not be one for a long time to come. You have the timecodes for the two sources for the effect and you have the VHS. If you are really persnickety you may have even written down some of the effect information as you made the effect. Adding a comment to the list like, "second keyframe is exactly 20 frames later" doesn't hurt.

This information is best delivered to the EDL itself with the Add Comment feature in the MC/AX. If you highlight a single shot in the sequence with the segment arrow, you can pull down the menu above the record window and request the Add Comment option. The comment shows up in the EDL as a little bit of extra text when you choose Show Comments in EDL Manager. You can search by comment in the sequence using Command-F. Think of it as locator information that makes it to the EDL. Now you can enter information about the sequence or the transitions.

Even if you can represent exactly the way an effect was going to happen, you cannot foresee how the online edit suite will be wired. The effect may call for three levels of dissolves, but the switcher only has two levels it can do simultaneously. Your effect may require that the editor create multiple layers of video and that is something the editors should know how to do quickly and efficiently with their own configuration of equipment. If you try to be too specific about the exact procedure for creating an effect, you are denying the online editor the chance to do what they do best.

## MULTIPLE LAYERS OF GRAPHICS AND VIDEO

The final insult in translating from nonlinear to linear is the way most EDLs represent multiple layers of video. They really don't. This means you need to reduce all multi-layered sequences to a series of single video track EDLs. The nested tracks must be broken out of track one and put into separate EDLs for each track in the nest. This provides you with the ability to run more than two video sources simultaneously in the online edit suite. Don't count on it unless you have researched the specific suite you will be using.

The online editor loads the video tracks as separate sequences into the edit controller and combines them. You can help the online editor figure out which tracks go where by the way the separate lists are numbered. Each edit is called an *event* and has a number. If the first layer of events ends at event 110, then all of the events for the second layer of video can start with event 200. The third layer can start at event 300, the fourth layer at event 400, and so on. The editor knows that when two events overlap in the EDL and if one has event number 50 and the other is event 210, then 210 is part of a multi-layered effect.

Any multi-layered sequence should be carefully discussed with the individual editor you will online with. Make sure they can even get close to what you are doing on the Avid. You may find they say, "Sure, we can do that—in the Flame suite!" Listen for the cash register sound when they want to put you into a special compositing suite. There may be no other choice but to go this route. On the other hand, a slight redesign and some extra goofing around may save you thousands. A freelancer using Adobe After Effects or another desktop compositing program may be the right answer if you have the time to render. There is a tradeoff between waiting for a short render in the expensive compositing suite and waiting until Monday for After Effects. With the use of render acceleration boards like ICE (Integrated Computing Engines) for Adobe After Effects, the render times on the desktop are becoming more competitive. It may make sense for you to get a cheaper online suite and arrive with all the effects already made so you just slug them in. Did I mention that you can't make changes in the online suite?

## SIMPLIFY THE EDL

If you try to make an EDL from a complex sequence with layers and nesting and you just load it all into the EDL Manager and press the Purée button (or Update) what comes out will not be very useful. You may get some error message as the system warns you that what you are trying to do is not very nice — usually some kind of a parsing error or a message that an area of the sequence is too complex to be represented. Make a copy of your sequence and start to simplify.

First, get rid of the nesting. Figure out what are the most important sources for that effect. What timecodes do you need? Keep just that source. If you absolutely have to have the timecodes from multiple sources for that effect, you should not have nested it in the first place. Yeah right, that assumes you knew you were going to an EDL. If you are going to an EDL, do not nest. Chapter 10 describes how to unnest. You enter the extra timecode and source information by hand as a Comment. In EDL Manager you choose which video track to use; start with V1.

If you know you are going to an EDL, don't use PICT files or imported graphics. An online suite may be able to use PICT files through a sophisticated character generator, but if you know you're going to an online room, all graphics should be fed to you on tape. The timecode numbers of the fill, matte, or backgrounds can go cleanly into your EDL with few questions asked. Also, if you use PICTs, it means additional work for the graphics department because they must output the PICT for you, then they have to make a fill and matte and lay it to tape for the online. If you have subtle variations in logos or graphics, the online editor can put the wrong graphic in since now you expect them to figure out which graphic on the graphics tape corresponds to your non-timecoded PICT source. Even if you use PICTs to simplify your edit or because you didn't know you were going to an EDL, I strongly recommend recutting those sections with the same graphics tape that is used for the online.

## SOUND LEVELS

How did you change your sound levels when you were mixing? If you used rubberbanding there is no problem. Those levels are not in the EDL, but the sound edits do not confuse the online editor either. If you used the older method of creating Add Edits, changing the levels, and adding dissolves to smooth out the level change, then you have some work to do. These Add Edits are real edits and the EDL Manager is smart enough to ignore them, unless you have added a dissolve. The EDL Manager puts an edit in the EDL and tells the online editor to dissolve to a B roll, a dub of the exact same audio source, just to change the audio level. This is confusing and annoying. You need to remove the dissolves to any audio transition that is just for a level change. This is actually pretty easy because you can go into the effect mode and lasso or Shift-click them and delete them all at once. This is a good reason to use rubberbanding.

For the really high-end jobs, the sound is usually sent out for audio sweetening and is laid into the online session in one big edit. For lower cost jobs, a digital cut is laid to tape (sometimes split tracks, sometimes stereo split) along with the low-resolution video. Then the online editor cuts his video on top of that, covering it. For these EDLs you should disable the audio tracks, creating a nice, clean video-only EDL. If you want to be safe, you can make one EDL with audio just in case they need to find the video for a sound bite that has been entirely covered up or if there is a problem with the sound from the Avid in a certain spot.

## SETTINGS

There still seems to be a lot of choices in the EDL Manager that haven't been described yet, so don't get nervous. Most of these functions are for streamlining

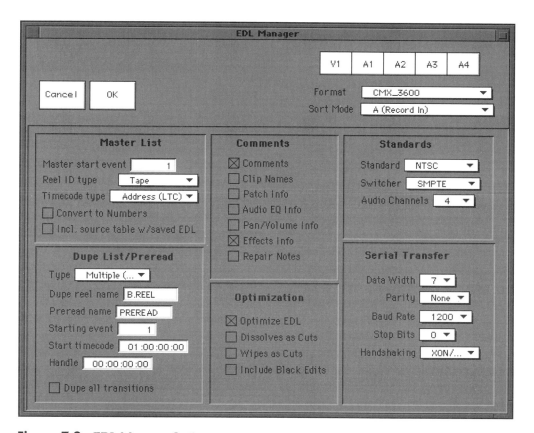

**Figure 7.2** EDL Manager Options

an auto-assemble and other more obscure situations. We will get to them at the end of the chapter, but for now let's go with the simple answers.

## SORT MODES

A sort mode is a way of ordering the events in an EDL so that they can be edited out of linear order. Linear order means that you load a tape into the deck, adjust the levels from the color bars at the beginning of the tape, fast forward to the spot on the tape where the shot is, and perform the edit. Then you load the tape for the next shot in the program and do the same thing. You do this whether the source tape is 20 minutes long or 2 hours long. There are a lot of repeated actions and wasted time shuttling the source tapes back and forth. Unfortunately, a linear order is the easiest to follow and the safest way to assemble a show and also be able to compensate for a mistake or make changes. You know immediately if a shot is missing or a sound bite is wrong because it does not make sense as you put it in place.

The time savings are significant if you can assemble a show out of linear order. It means you only need to load a source tape once, do all that color bars set-up, and skip through the show, dropping shots wherever they are needed from that source reel. Then put the tape away. You might have to bring it back for effects later, but if your show is just cuts and dissolves, you have made serious savings in time and money. But imagine making a change that makes the show just a few seconds longer. If you have been inserting shots all over the master tape out of order, this is pretty serious. How can you ripple that change, push down all the edits after it in the program, if there are already 100 shots laid down after the change? You can't really; if you are working with digital tape you can clone the master, make an exact digital copy, but if it is an hour-long program and you are half way done, then kiss goodbye a good 45 minutes of set-up and dubbing time. Did I mention that you can't make changes in online?

This kind of penalty for being wrong or making changes at the last minute tends to frighten even very brave online editors. For this reason they want to be the person that makes the decision how to sort the list. When asked, they inevitably tell you to use the A sort mode because that is the linear mode and they sort it themselves in their edit controller to make it a B, C, D, E, or S mode.

Let's do a quick recap of what the different modes are so you can discuss them intelligently at cocktail parties. B mode is for long-format shows where the source tapes are shorter than the master tape. The source tape fast forwards and rewinds and finds shots that go in a linear order on the master. This is easier to follow because the master moves forward while a specific source tape is being used. As soon as a new tape is loaded, the master rewinds to the first place the new tape is used and starts to move forward again until all the shots from that source are inserted on the master.

A C mode list is for when the source tapes are longer than the master tape. You may have a 90-minute film transfer for a 30-second commercial. (I hesitate to call them "spots" anymore since in the UK that means pimples!) Here, the source tape always moves forward in a linear way and the master tape flails around rewinding and fast forwarding to assemble out of linear order. This is potentially faster than B mode but harder to figure out what the heck is going on. Very disconcerting for clients who should just look away if they are feeling queasy.

D and E modes are like B and C modes except that all the dissolves are saved for the end. This works only if you have time base correctors (TBC) that have memory for individual tape settings. The editor must match a frame perfectly from something that he adjusted this morning or yesterday so that you don't see where the pickup point is to start the dissolve. This gets used more in PAL countries where they don't worry about hue adjustment, unless it's terribly wrong to start with!

S mode is useful for getting a list that will be used for a retransfer from film. If you have been cutting with a one light or a best light, inexpensive film transfer without shot-by-shot color correction, then you want to go back and retransfer the shots that were actually used. S mode is the order that the shots came on the original film reels and allows the colorist to quickly identify only the used shots for the final color correction.

## DUPE REELS

After choosing the format and the sort mode, it is time to think about dupe reels. What are you going to do when you need to dissolve or wipe? If you haven't experienced the difficulty of dissolving between two shots on the same tape then you are about to. With a traditional analog tape suite you must physically copy the second shot onto a dupe reel so that you can roll two tapes at the same time. Seems kind of old-fashioned and quaint now, but is still the necessary evil in many suites across the world. The biggest change to this procedure was *preread*. Someone figured out that since the digital format tape was just outputting digital information, short amounts could easily be held in a buffer, or temporary memory in the record tape deck. During the preroll, the master record machine loads the digital images of the shot already on the master tape into the memory and plays it back as a source deck. With the last shot held in memory, the master tape switches from being a source and records the dissolve with the new, incoming shot.

Some editors also use a kind of non-destructive preread called *autocaching* which holds the end frames of a shot for dissolves and various layers

**Figure 7.3** Dupe Reel List

of multi-layered work in a cache that uses a digital disk recorder (DDR) or a digital tape. The edit controller figures out what needs to be cached for later use in a dissolve or effect. Then it automatically lays it on the DDR or digital tape and when the tape with the other half of the dissolve is put into a deck, the controller performs the dissolve between the tape and the auto-cached piece of video. This works better than preread because prereads are destructive, meaning you are committed to that edit permanently because the last image is recorded over itself when the recorder becomes a source. You only get one chance to do a preread correctly and then you must re-edit the previous shot. Auto-caching uses a preread EDL to create the auto-caches that can be executed and re-executed if needed.

Most of the choices that deal with how a tape is dissolved are refinements on these basic requirements for either a dupe reel or preread. Under certain conditions, you can significantly reduce the amount of tape loading and shuttling, but most of the time you choose the simplest technique. The online editor has a very specific requirement and again, like with sort modes, can modify the simplest format into something they can use best.

The simplest, best answer for most circumstances is the Multiple B Rolls option. It is also the least clever, but with more sophisticated editing systems out there, editors can modify this list to speed up the assembly. The Multiple B Rolls option assumes that there is a separate B roll or dub of every tape in the show. This is probably not true unless you have a very small amount of material or have made simultaneous duplicates during a film transfer. It is more likely that the online editor looks at the list, sees that there is a dub required, and hunts down the original tape. The timecodes on the B roll are the same as on the original tape and the name of the B roll is a variation of the original tape name. Finding the shot is easy because it has a "B" after the tape name or the tape number is incremented by 500, depending on the type of list format. The editor can then just dub that shot when it is needed or change it over to preread in the edit controller.

## Making a Special Dupe Reel

The other B roll choices allow the editor to prepare before the online session by taking all the shots that need to be dubbed and putting them onto one reel. This keeps the tape loading down to a minimum and allows you to move faster with fewer decks configured (and charged for).

The most dangerous dupe reel choice of all is the One, New Timecodes option, although it is extremely popular in Europe. This requires that you make two lists, one for the master tape and another for the tape where all the B shots will be dubbed. This new tape has perfect timecode because you are ignoring the original timecodes of the source tapes and starting all over again. You take each tape and dub one after the other onto this new reel, which starts with timecode that begins at 1:00:00:00 (although you can change this). This means that the link between the original source tape and the dupe reel will be completely broken as soon as the dupe reel is made. The dupe reel name, which defaults to B.REEL, shows up in the list every time a shot needs to be dubbed for an effect. You cannot look at the EDL dupe reel name and figure out where the shots came from originally. Many people choose One, New Timecodes and then do not make the dupe reel list. This is a big mistake. When the online editor turns around and asks, "OK where is this B.REEL?" you really should have another list to hand over. That reel needs to be assembled at the beginning of the session.

Since it is assembled before the editing really starts, changes to the B.REEL are also very difficult. You must make a second list for the dupe reel! If you must make a dupe reel then One, Same Timecode is better. If you forget to make the list for the new dupe reel, at least the original timecodes are in the list and, if the timecode hour matches the tape name, finding the original is a snap.

Another dangerous choice for the unsure is None. This gives you a list that most edit controllers can not read, which at first seems like a rather bad choice. Not so, if you are going to another NLE or an effects box like the Discreet Logic Flame or Quantel Henry. These systems do not need to dub a tape in order to dissolve because, once digitized, individual shots are divorced from the physical restrictions of tape. Loading the list into the NLE systems forces it to see the tape 001 and another tape named 001B. Is this the same tape? There is no way to know for sure, so the NLE system makes the safe but dumb assumption that 001 and 001B are two distinct tapes. When you are digitizing this material, the system asks for tape 001 and then proceeds through this tape to the end and grabs all the shots that are needed for the sequence. Then the NLE system asks for tape 001B, rewinds and does the same thing again with the same tape! This is not efficient and, at those room rates, you need to be efficient. The None choice shows just one tape, 001, and the NLE system only requests that tape once.

## PRINTING THE LIST

When you print out the list you should make two versions. Once you are ready to print the EDL, think about who will be using the printout. The person supervising the online edit certainly needs it in front of them. They may not be the most list-savvy person in the production so there are ways to make the list more readable. Include comments in the list when you are about to print. Use comments like "clip name" and the source table, which is a list of all the sources needed for the online assembly. Then print a stripped-down version for the online editor. Again, ask if there is any special information the editor wants, like audio patch information or repair notes. I know some offline editors are reluctant to have repair notes show up in their list because it looks like they did something wrong, but it may just be a case of the list being too complex and not being able to represent it correctly. If you have any repair notes then the online editor should see them.

## THE SOURCE TABLE

The important thing about the source table is that if the EDL Manager found a tape with a name that was too long and had to shorten it or found a duplicate tape name from another project, it has to change that tape name in the list. The source table tells you that since you had a tape 001 from project X and a tape 001 from project Y, that, with a Sony format, EDL Manager changed the second tape name to 999. A CMX list changes a second tape 001 to 253. According to the system these are two different tapes! If you don't have that information then you

**Figure 7.4** Source Table

could get very lost. Never attach a source table to the EDL and always print it out separately to take with you.

## PAL LISTS

If you have a PAL project, be aware that the very first time you open EDL manager, it defaults to NTSC. You can change this to PAL under the Standards setting and it will stay that way. PAL lists become complicated when you are working in film and want to go back to the negative because of the two different methods of dealing with the difference between 24 fps and 25 fps. Sometimes you slow down the audio and sometimes you include extra frames in the video (that pesky 25th frame). The PAL Method 1 is for film projects in which video and audio are transferred together at 25 fps. PAL Method 2 is for when video and audio are digitized separately and synched in the Avid. If you have a video project without any matching back to film, then choose plain vanilla video PAL.

## MATCHBACK

Matchback is a method preferred by low-budget films or projects that are primarily video but will be projected at sometime in the distant future. This is an option for both Media Composer and Avid Xpress that will output a 30 or 25 fps project as a 24 fps cut list. Because film transfer to video adds extra frames and there is no way for the video-based project to keep track of the A frame, then every outpoint in a matchback project can only be accurate to plus or minus a video frame in NTSC. For many people, depending on the material, this is quite acceptable compared to the cost of conforming a negative first and then transferring to video or using a Film Composer. They get a video project at a high, two-field AVR that they can shop around or distribute and when the funding begins to flow in, they can get the negative cut.

When preparing for online, you need to simplify a copy of the sequence before you make the list based on feedback from the online editor. Then find the right edit controller format, format the floppy for the needs of the production house, and cover yourself by making what seems like about a dozen more versions. Print the list twice. Make a VHS copy of the sequence that you can control yourself in the suite and don't forget to bring the source tapes!

# 8

# Finishing on the Avid

One of the most exciting aspects of the digital nonlinear technology is the promise of finishing a project completely digitally. The goal of never going to tape is a reality today for many producers, especially for newsgathering. The post-production equivalent is the ability to input everything into one system and free yourself from having to go to any other facility to output a high-quality, professional finished product. As we discussed earlier, this means mastering new skills and taking on the responsibilities that other specialists have taken care of in the past. Specifically, it means learning to set your own video and audio levels. It also means keeping the entire workflow in mind at even the earliest stages and scheduling enough time for all the details at the end of the project.

First, let's look at the specific workflow involved in finishing on MC/AX. If you have been digitizing at high resolution throughout the project, you can output to tape and be done with it. All your levels should get close scrutiny at this stage because you are the last person to run the project through a quality check. This is a serious responsibility and you should take nothing for granted.

If you have been working at low resolution throughout the project, then there are quite a few more steps. In the simplest project, you redigitize the shots in the sequence, add in the high-resolution graphics, render the effects, and digital cut to tape. If your sequence is that easy and straightforward, this process will be simple and efficient. If you have a long, complicated sequence or some not-so-simple demands from the client, then you need the extra power and flexibility that the techniques described in this chapter give you. Some of them are basic and intuitive and some are pretty complex. I invented some of the more complex techniques in response to real-world complications. These techniques have been previously published on the Internet, given out at seminars, and faxed by Customer Support. Simple is good, but flexible is better.

Finishing is one situation where Media Composer has an advantage over Avid Xpress. Avid Xpress is missing one important feature to speed the offline to online process: decompose. Although this sounds like it would smell bad, think of it as a natural step in the life cycle of the project. It gets you across the detailed transition from offline to online. You can, of course, go from low resolution to high resolution without decompose, but it is not as fast, flexible, or efficient. We will describe this feature in its proper place in the overall process.

## PICKING THE RIGHT AVR

Look at the total amount of material that you can realistically put onto the storage available to your system. At this point in the technology, the limitations have much to do with the Macintosh, your type of the SCSI accelerator card (single ended or differential), and the length of SCSI cable allowed. The rules are changing pretty fast and, with the widespread acceptance of fiber channel, you will see some pretty amazing drive farms. So for offline, use the lowest AVR you can stand without squinting; think of the low AVR as a new form of impressionism! If you are confident of the quality of the footage, for example, a film transfer from an excellent director of photography, all you really need to see is boom shadows, lip synch, and distracting backgrounds. However, the client may determine the quality and leave you no choice but to figure out how to get all your media on your drive at that (higher) AVR. Some people, especially those not use to working with compressed images, get very nervous when viewing low AVRs and just don't want to put up with the lower quality. They must have the better-looking image throughout the project.

With the NuVista+ video card you can not mix AVRs at all. Once you choose an AVR, stick with it throughout the offline or online part of the project. With the Avid Broadcast Video Board (ABVB), you can mix single-field AVRs (2s, 3s, 4s, 6s, 8s, 9s) with other single field AVRs, multicam resolutions (2m, 3m, 4m, 6m) with other multicam resolutions and two-field AVRs (12, 70, 71, 75, 77) with other two-field AVRs. This means that with the ABVB, you can digitize all your timecoded footage at AVR 3s and import all your graphics at AVR 9s. Many people like the option of AVR 12, which is a low-resolution, two-field AVR. Using AVR 12 at the offline stage allows you to digitize all the timecoded footage at low resolution, but import all the graphics at AVR 75 or 77. However, because it is a two-field AVR, AVR 12 takes up more storage space and does not look quite as good as a similar low-resolution, single-field AVR. This is because the second field in AVR 12 is not to gain you better resolution, but to be compatible with the finishing resolutions. It saves time at the last stage because all your graphics will already be at AVR 75 or 77.

Many film projects are using the multicam resolutions because they take up less space than similar "s" resolutions. If you are using the "m" resolutions, make sure to turn off or disable your 3-D effects. Do this by quitting the MC/AX and relaunching while holding down the "F" and "X" keys. Your sound will always be CD quality or higher no matter what AVR you choose, so there is no need for redigitizing there.

After you pick the right resolution for your project, the digitizing can begin. It is very important that you set the levels for each tape at this stage. You can save these level settings attached to the tape name and they will be recalled automatically at the batch redigitizing stage of the online. Get the levels correct now and you save huge amounts of time later when the deadline is tight and time is potentially more expensive. If there are multiple, radically different levels on the same tape, you may end up creating several settings for that tape. Give that tape multiple tape names, a new name for each setting. For instance, you may have tape 005 and it is primarily exteriors. Then there is a change to an interior set up at the end of the tape that requires a different set of video levels to compensate. You might pop the tape out of the deck, reinsert it, and rename the tape 005IN or something similar. Change the level setting to compensate for the interior and label the outside of tape to reflect that the system now thinks there are two tapes.

After the levels are set correctly for the video, make sure that the audio levels are set correctly too, and watch them closely. Always keep in mind that you can use this audio as the finished mix. If you are going to end up on digital tape, you can digitize at 48 kHz and go digitally to tape (AES/EBU) at the end. This also means that if you choose 48 kHz now you won't be able to move the project halfway through the job to a lower end system like a Media Composer or Avid Xpress without the 48 kHz option. There are DAT decks that accept 44.1 kHz so they may be a good alternative if you have one of these lower end systems and want to work with digital audio. The Avid Xpress and lower end Media Composers input and output S/PDIF (Sony Phillips Digital Interface Format), which may not be available on some professional level DAT decks. They use the AES/EBU standard connection.

If the audio goes in to the system right and you can mix the sound as well as needed in the Avid suite, then you will save a significant amount of money. The new Audio Suite sound plug-ins give you an excellent new tool for the finishing touches or special effects. On the other hand, professional audio mixers can be worth their weight in gold on the right job. If you digitize the levels right, you may be able to hand those audio media files directly to the audio mixers as AIFC (a variation of AIFF) or SD2 files with OMFI sequence information for Pro Tools, AudioVision, or other digital audio workstations (DAW).

When graphics and animations are created, you import them and cut them into the show as needed. The best scenario is to import them at the finishing quality of AVR 77. If not, you will need to reimport them later when you get to the online stage. Reimporting can be tedious if the graphics have alpha channels. I will outline a method later in this chapter that I devised to batch import graphics and animations without alpha channels.

Edit the sequence together until you get the final authority to sign off on it. Many people call this *picture lock* and assume that nothing is going to change after everyone sees this final version. (Don't ever believe this, but it is a nice idea.) In the process of getting final sign-off, you need to get this low-resolution version as close to the final high-resolution version as possible. That entails, in many cases, creating all the effects at this offline stage. If straight cuts are more than enough for your production, consider yourself lucky, and in the minority. You may have a specific look that has been mandated by the designers and it must be accurate down to the kerning and leading of every letter of every title. This, too, must be created in the offline so that there are no chances for them to say, "I didn't know it was going to look like that!" and start redesigning while the clock ticks. Everyone's approval in the offline stage is important.

This chapter will discuss each of these steps in detail. The important factors are quality control and the speed of graphics replacement and rendering. The video and audio levels must be monitored at every step to make sure they conform to broadcast specifications. Most people render too much. If you have overnight, then this is not an issue, but if you are trying to make an air date then saving even a few minutes becomes crucial. Chapter 10 discusses strategies for simplifying effects sequences and reducing rendering.

## NECESSARY EQUIPMENT

Your Avid suite is now an online suite if you plan to finish there. By purchasing the highest AVR and the fastest drives you have not completed this transformation! Online suites have some important components that must be present. There must be a high-quality third monitor to view reference video. This must be an engineering-quality monitor that can be adjusted through standard industry adjustment procedures like using the blue bars setting. The monitor must hold that calibration in a stable fashion over time. This generally means more money for the monitor, but if you ever get into a dispute about color or brightness, this monitor is your absolute. It is the end of argument and the last word on "what it really looks like." Don't put your faith in cheap equipment—this finished image is your reputation!

Some people like to have a consumer-type monitor in the room as a low-end reference to "see what it will look like at home." Be careful of this since it is very hard in NTSC (Never Twice the Same Color) to ever get certain colors to look exactly the same on both monitors. When your client insists that the yellow on the low-end Gold Star monitor must match your high-end Sony monitor, you have a frustrating no-win situation. This is why many online tape suites have only one color monitor and everything else is black and white.

There are very precise color-measuring devices that, in the right hands, allow you to adjust monitors precisely. After such an adjustment, the monitor should be left on all the time to minimize the drifting that occurs with warming up and cooling down. If you need to adjust a color monitor, always wait until it has been on for several minutes, longer for older monitors.

It is also nearly impossible to get the computer monitors to match up to the video monitors. They are two very different types of monitors and, even though you can adjust the computer monitors, you cannot adjust hue through hardware adjustments. There are color-matching extensions that give you more control than the front panel buttons. Many professional graphics programs include these gamma-adjusting extensions, but you will never get the monitors to match completely. Do not let the client pick a final color on the computer monitor. They must have final approval on the well-adjusted, carefully lit, high-end video monitor. And yes, it will look different at home.

You will need external waveform and vectorscopes as an independent reference for all signals. Use it to monitor input, output, and dubs. If you can set up a patch bay or router, everything should go through these "scopes."

More and more people are deciding that having a color corrector in the suite gives them the extra protection and ability to deal with sources that have difficult color problems. It also gives you an extra tool to make sure that sources shot in very different locations really match up when they are next to each other. Since there is no hue adjustment for component NTSC and no adjustments at all for serial digital input or PAL, the color corrector gives you the level of control you occasionally need. It can also give you some illegal levels if you are not careful! If that black looks really rich and the shadows have that deep dramatic look you really want, check that the black level never goes too low.

You need studio quality speakers and they must be mounted far enough apart so that you can be critically listening for stereo separation. If you can afford it, get a good compressor/limiter. This takes your audio levels and gently compresses the loudest parts of the sound so that they don't distort. This means you can have your overall sound louder and not worry about the occasional spike in sound level. You can use it both during the digitizing and the output. Most Top 40 radio stations do this to everything they broadcast, which allows them an even volume level: loud. If you are sending your audio to an audio expert to finish, don't compress it. Leave the sound alone except to adjust levels and make sure nothing is distorted.

## CHOOSING THE INPUT

It may seem obvious that you always want to choose the highest quality input possible, but it is not always obvious exactly how to go about making the right choice. Let's start at the worst quality choice possible: composite. Composite video has the color information and the luminance information all crammed into one, antique signal. This signal has changed little since the early days of television. The equipment that generates, carries, amplifies, and broadcasts the composite signal has improved significantly, but not the signal itself. The PAL composite signal is better than the NTSC composite signal because it has more scan lines and the color is more stable, but both can still be improved. Even if you must use a composite signal, you can avoid digitizing into the composite inputs of the MC/AX.

The main problem with composite is that it must be decoded to separate the color information from the luminance. The decoding process is hit or miss unless you have dedicated hardware doing the separating of the signal: a decoder. Separating luminance information from chroma in the composite signal is as difficult as if I told you the number 12 and you had to tell me the exact two numbers I used to get that number. And the two numbers change all the time. You probably would get it wrong occasionally, maybe even most of the time. That is how a decoder must figure out where the luminance stops and the chrominance begins in a signal where they are combined. Good decoders actually change how they decode over time and have expensive, adaptive comb filters. Frankly, good decoders do a better job of decoding than simpler luma/chroma separation methods. If you are relying on composite sources for online quality, it is well worth your money to invest in a composite-to-component decoder. Only buy the types of converters that have a minimal delay since some models use a frame buffer and put you out of sync by one frame. If you only need to use a composite source occasionally, you might be better off sending the composite signal to the composite input of a component tape deck, like a Betacam SP, and using the component output of the Betacam SP to go into the video board for digitizing.

Some decoders even have an input port for S-Video and, if possible, you should use the decoder for that too. Take the S-Video output of Hi-8 or S-VHS and decode that to component for digitizing. S-Video, sometimes called Y/C, is a better quality signal than composite because it separates the luminance (Y) and the chrominance (C) information from the beginning. But the chroma is just like composite chroma and has all the same problems with crawling and bleeding. You have more bandwidth for the detail in the luminance part of the signal and the chroma does not have to be separated, but it is still not as good as real Betacam component. Middle range and low-end Betacam SP decks have an S-Video in and out. As with the composite signal, you can run the S-Video signal through the Betacam SP deck and input component into the video board by using a Betacam SP as the decoder. Make sure that all tape decks are using the same blackburst signal for a reference to make this work.

Then there is true Betacam component, more technically known as Y, R-Y, B-Y. The Y stands for the luminance and holds all the detail of the image. The R-Y is the red component of the image with all the luminance information eliminated and B-Y is the same for blue. Together, they can create all the colors in the YRB color space (sometimes called YUV), what you can normally reproduce with Betacam SP images. The ABVB (Avid Broadcast Video Board) accepts Y, R-Y, B-Y, and composite input, but not S-Video. The NuVista+ video board accepts composite, S-Video, and RGB. RGB is the type of component signal the computer uses internally when it processes the signal. For the NuVista+ to accept Betacam SP component it must be transcoded from Y, R-Y, B-Y to RGB. This requires a special box called a *transcoder* to change the signal from one type of component to another for both input and output. Getting a transcoder for the NuVista+ is really worth the money if you are finishing.

When you get into the digital formats like D1, D5, and Digital Betacam, the best way to go into MC/AX is serial digital and for this you need a special serial digital ABVB. Serial digital is also a good way to input DVC Pro since it also has a serial digital output. You may see other alternatives for DVC Pro in the near future that will take advantage of the faster than real-time transfer rates, but it will probably happen in news first. Serial digital input means that you don't have to convert from digital to analog and then back to digital again when working with digital tape sources. The digital information is copied over to the Avid just like copying digital information between two hard drives. This produces superior quality, but if you have a mix of sources, like the occasional source tape from Betacam SP, then you must convert from analog to serial digital with a stand-alone hardware converter. There is only one kind of video input with the serial digital ABVB, serial digital. When you output from a serial digital ABVB there is a monitor-quality composite output besides the serial digital output. This composite output should only be used for analog video monitors. To truly see how good the signal looks, you should be using a digital monitor anyway. You can probably use the composite output for a VHS preview, but not for finished output. Why waste all that quality?

With the release of version 7.0, the Media Composer and the Avid Xpress will be compatible with EditCam media files, the field camera developed with Ikegami to record directly to disk. This new technology is exciting and the idea of not having to digitize at all has obvious benefits in both time and money. You will also be able to add comments, do rough cuts, create bins, and many other traditional post-production tasks while still in the field. Connect the EditCam field pack to the editing system and import into your project.

## SETTING LEVELS

In the perfect world no tape should be digitized without being setup to broadcast specifications. This means looking at the tape carefully, both through the Avid internal waveform and vectorscopes and external scopes. Both types of scopes serve important purposes and both are necessary. Don't set up an online suite with scopes! You use them when you digitize and when you are adjusting color effects, imported graphics, and dubbing. Getting a show rejected by broadcast engineers or a laser disc facility because the levels are outside standard limitations is costly and embarrassing.

Video levels in the ABVB go from .74 IRE to 108.4 IRE in NTSC and .249 volts to 1.063 volts in PAL. The ends of this range are called *footroom* and *headroom* and is important for reasons having to do with graphics using superblack (levels below normal black) and the overall desire to keep the signal from being clipped off if it is just a little outside of broadcast specifications. Every end use has its own set of standards. Public broadcasting in the U.S. has slightly more conservative standards than other broadcast outlets. Laser disk manufacturers have different specifications and may require different kinds and amounts of color bars and tone on the master tape. Find out what your requirements are and use the scopes throughout the project to conform to those specifications.

Many people memorize which scan line to use for each of the level adjustments using the Avid scopes, but this will fail you the first time you are faced with different kinds of color bars. Instead, understand why you are looking at a specific part of the screen and what information the color bars are showing you there. In order to see the perfect color bar pattern, open the waveform or vectorscope and hold down the option key.

There are several different kinds of color bars depending on where the tape originated. The most common color bars for NTSC purposes are SMPTE bars, which have several different sections or fields for different adjustments. The top section is used to adjust color on the vectorscope. The very bottom section is used to adjust brightness since it contains 100 IRE white. It is also used to adjust black level, or setup, because it contains 3 kinds of black, 3.5 IRE, 7.5 IRE, and 11.5 IRE. The middle value should be on the dotted 7.5 line. The other two values of black in NTSC are called *pluges* (Picture Line Up Generating Equipment) and are meant to be used to adjust your video monitor brightness. When you adjust brightness level on the monitor you should see no difference between the pluge

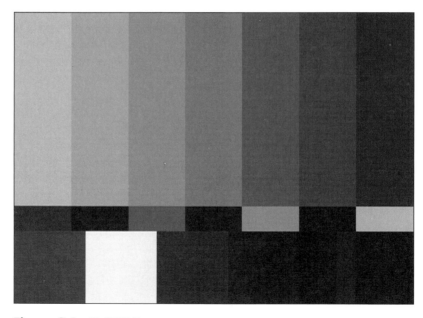

**Figure 8.1** SMPTE Bars

at 3.5 IRE (the one on the left of the group) and the setup at 7.5 IRE (the middle bar). In PAL you can set the monitor to a cross pulse or underscan setting and compare the black of the now-visible blanking signal to the black in the color bars. When you are adjusting the video input levels for gain (brightness) and setup, you must move the scan line being monitored to the bottom of the frame until it looks like this on the waveform (see Figure 8.3).

The second most common type of color bars are full field bars, which just have straight bars from top to bottom. These are simpler to use, but they tell you less. You can put the scan line just about anywhere in the image (except the very top and bottom) to get a proper reading for brightness and color. The important thing to determine about full field bars is whether the white bar is 100% or 75% white and whether the chroma saturation is 100% or 75%. Preferably, you should use the full

**Figure 8.2** Waveform Monitor and Vectorscope Showing Perfect Chroma (NTSC)

**Figure 8.3** Waveform Monitor and Vectorscope Showing Perfect Luminance (NTSC)

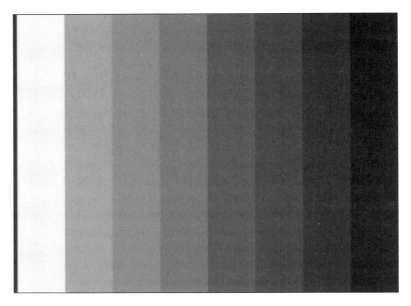

**Figure 8.4**   Full Field Bars

field bars that have 100% white and 75% chroma saturation. These are commonly referred to, rather vaguely, as 75% bars and more specifically as 100/75 bars or 100/0/75/0 bars. If the white bar is at 100%, then set the white level to 100 IRE or 1 volt (PAL). If the white bar is 75% white, then in NTSC, set the white level at the dotted line at 77 IRE. The math is strange in NTSC because of the 7.5 IRE black level so 75% becomes 77 IRE. PAL has no setup so 75% white should be at .75 volts.

In NTSC there is hue adjustment with a composite signal. You can move the color wheel around until the points of the vector signal line up in the little boxes of the vectorscope. With component input, you can only have more of B-Y (blue minus luminance) or R-Y (red minus luminance). More or less of these colors puts the vector points in the boxes. If the points don't fit in the boxes, you need to use a component color corrector during the digitizing or apply a color effect to the clip after it has been digitized. You can just set every thing to the factory set defaults levels of the video boards and send the finished master out for tape-to-tape color correction! That is not a real alternative for most people and besides, if you clip the signal by letting it get too bright or crush the signal by letting it get too dark, any color correction afterwards will only mask the damage. Get the levels right the first time and any further manipulation will look better.

A serial digital signal has no adjustments for video levels within the Avid Video Tool. You can adjust the levels from on the tape deck itself or use a color effect after the image has been digitized. If you are coming from analog sources, you need a digital converter and maybe even a color corrector. One popular digital color corrector has a "make legal" button. If only the rest of life was that easy!

## WHERE TO ADJUST LEVELS

There are several schools of thought about the best place to set input levels: the deck TBC or the Avid. Frankly, according to the engineers who designed the ABVB, they are both good, with certain caveats. Let's look at both and you can decide which one is better for the way you work.

### Output

Both methods require first setting the output levels of the ABVB. The ABVB has a separate tool for setting output levels; the NuVista+ does not. If you do not set the output levels before you set the input levels, you may be masking a problem. In audio, if you turn down the volume of a signal that is already distorted you get quiet distortion. In video you get a darker image that has already lost all its detail. So make sure that the output levels are not affecting the signal at all. There is a very easy way to do this, but it requires using external scopes and a built-in test signal. If you (foolishly) do not own external scopes, make sure the levels in the Output Tool are in the green, preset position. Otherwise use the toggle arrows at the bottom of the Output Tool (more choices in 6.$x$/1.$x$) and choose one of the color bar patterns under Test Patterns. These are uncompressed PICT files that are designed to show digital video levels (ITU-R 601) and should register as the most beautiful and perfect pattern you have ever seen on your waveform and vectorscopes.

When you are outputting this perfect signal and looking at the external scopes, you may occasionally see a slight discrepancy. This may be because of other devices that are in the signal path between the video board output and the scope. You can eliminate this guesswork by putting the scopes immediately at the output of the video board. If this looks good, you have a problem down the line, perhaps with a distribution amplifier (DA). A DA is meant to keep the signal from fading over long lengths of cable or for allowing you to split the signal up into many outputs for routers or multiple decks. The DA must be adjusted as well!

If the signal is still a little off and you have eliminated the obvious, then you may have to adjust the output levels. If you do this, save them as settings, put them in the Site Settings and call them "Default." That way, any new project will have access to these settings and can use them instead of the factory presets.

### Input

The first method makes the most sense from the point of view of editors who have been setting all external devices to unity. In other words, before they send a signal to the DVE (digital video effects device), they adjust the input levels of the DVE so that all signals coming in look exactly the same coming out. That is called *unity*. What goes in is exactly what comes out. This method is also good for editors used to working with a complicated video pathway through a large online suite, routing system, and machine room.

Now feed the Avid a color bar signal, preferably from a color bar generator, but any tape that is known to have good color bars will do. You may want to make a tape of just color bars and 1 kHz audio tone and keep it around as your reference tape. The MC/AX outputs perfect color bars and tone from the Output Tool and the Audio Tool and you can just roll tape for five or ten minutes and record it. Set the output of any playback deck so that it looks perfect on the external scopes and then feed the signal to the Avid. When you feed the bars to the input of the video board and set the levels so they are perfect on MC/AX's internal scopes, they should look perfect on the external scopes on the output of the MC/AX too. Save these tape settings and call them "Unity" or "Default." Save these settings somewhere safe and drag them to your Site settings. You now have standard default settings for both input and output. This is what should be used with all tapes when digitizing.

Now you adjust the output levels of the video deck using the time base corrector (TBC) to adjust for the differences in levels between tapes. You will never touch the settings on the Avid. This way you know the Avid is not introducing any "noise" or affecting the signal level in any way.

There is one big drawback to this method: There is no way to save the settings on the deck while there is a place to save the settings on the Avid. Some TBCs have a memory setting, but it is up to you to recall them by hand when you load each tape. What if you want to use the same tape two weeks later and you want the shot to exactly match the same shot digitized at the beginning of the project? You had better hope that you set everything perfectly up to bars and then did not make any further adjustments, because bars at the head of the tape are the only reference you have. Chances are, though, that you made minor adjustments after you set up the deck to bars because the picture didn't look just right. Unless you can reproduce that little level tweak by eye, you will have mismatching shots. It is much better to use the ability to save those levels and call them up automatically and you can, if you use a different method.

If you have not blown out or crushed a video signal while inputting, you can correct for any little adjustment in the Avid Video Tool without adding any extra noise to the signal. This is because the signal is digital and as long as it stays within a particular range (not distorted), it can be moved around without any further degradation. In this scenario you want the deck TBC to be always at the preset level. Throw all the video level switches on the deck into preset and then make the adjustments with the Avid Video Tool. Save the level settings based on tape name. Then, when you batch digitize, the settings for the tape are recalled automatically. If you have 100 tapes, this is a lifesaver. You set the levels once during the low-resolution offline stage and when you batch to a higher resolution, and are in a little more of a hurry, the levels are right on without any further intervention by the digitizer. This method works as long as both the offline and the online are done on the same computer.

If you are adjusting the deck, look into a remote control panel for the deck's TBC controls so you can adjust quickly on the fly. You really can't count on the settings for the color bars being the absolute perfect setting for everything on the tape. So even if you are working from the Avid setting, you may still want to tweak the deck TBC occasionally. The problem with taking a deck's TBC out of preset is that it jumps to wherever the knob was set the last time you used it. Then you have to quickly tweak the levels to get them back close to the preset setting. This makes it hard to do just a little adjustment, for instance, while the deck is prerolling and you see something that is just a little off. A good method, although it is an extra step, is to put the deck in preset, set up the Avid Video Tool Input levels, and save them. Then, while watching the external scopes, take the deck out of preset and manually adjust the levels on the TBC so that they look exactly like they are in preset. You are using the power of the reference point, preset, to get everything right, then allowing yourself to make minor adjustments based on those original, perfect levels.

## VERTICAL SHIFTING

Sometimes the entire image can shift down vertically and show you a black bar at the top of the frame and flashing little white lines. What you are seeing is VITC (vertical interval timecode), a second type of timecode that is recorded as part of the video in an otherwise unused part of the signal. You will see this problem quickly if you have set your video monitor on underscan. The source and record monitors in the Avid show you this after you have digitized, but ideally you want to catch this problem right away because it is so easy to fix.

This alternate type of timecode, VITC, is useful when you are moving the tape very slowly or very fast and the main form of timecode, LTC (longitudinal timecode), doesn't work as accurately because it is an audio signal. You know what happens to audio when you slow it way down; it becomes difficult to understand, and LTC is no different. Instead of guessing for the right timecode number at slow speeds, interpolating from the last well-read bit of timecode, a deck controller looks for the VITC. VITC also allows you to record extra pieces of information like tape name or film information that can be read by some timecode readers, like Media Reader and even burned-in to the image while digitizing.

VITC is really helpful until someone sets their record deck wrong and the LTC and VITC have different numbers. You see this if you are reading the correct number when the tape is playing and a ridiculous number when the tape is in pause. In that case, you want to go under the front panel of the video deck and change the switch to force it to read LTC only. Imagine what this would do to your logs if you didn't catch it!

Back to the vertical-shifting problem: What is happening is that the video deck is being put into an edit mode in order for MC/AX to control it. When a deck goes into an edit mode, it changes where it looks for a reference signal for its internal timing. The video deck changes from using the blackburst it is usually being fed from the blackburst generator for reference and instead looks to the incoming video signal. Many people are not feeding the deck a stable signal to the video inputs when they are digitizing. If there is no stable signal, it shifts the playback of the picture down several scan lines and reveals the otherwise hidden VITC. To prevent this, make sure the deck does not really go into the edit mode, always provide a stable video input to the deck (like a blackburst generator), or use a playback-only video deck.

The high-end series of record decks have a record inhibit switch that keeps the red light above the record button on the deck always on. This is an easy and fast answer until you want to make a digital cut. If you don't turn the record inhibit switch back off, MC/AX gives you an error as it attempts to cue up and make the edit. It may even ask you if your deck is in the record mode! Another answer takes a bit more preparation: Make sure that every tape you put into the deck has its record inhibit tab in the inhibit position. Betacam SP tapes and higher formats have a little red switch or a screw on the bottom of the tape cassette and 3/4" tapes require a little red plastic tab to be filling a hole in the bottom of the cassette. If you don't have the plastic tab for the 3/4" tapes to fill the hole then, as with most things in life, gaffers tape does the trick. Open-reel formats, like 1", have no protection on the tape itself so you need to rely on the record inhibit switch on the deck. Confirming that the record inhibit tab is in the inhibit position should be done by every tape operator, every assistant, and every logger in the world as standard operating procedure. A quick elbow in the wrong place with an unprotected tape in a record deck and you have a big ugly hole in your master source tape. Try explaining that to the producer!

Once you have inhibited the video deck from being able to make an edit on the source tape, the image correctly repositions itself vertically. If this does not work, you have a more serious blackburst problem and you need to evaluate that all pieces of equipment in the room are getting the same blackburst reference from the same blackburst generator. If you are working with a transcoder, this is probably the culprit. The blackburst reference input to the NuVista+ video board is the composite video in. If you are working almost exclusively with component or S-Video signals with a NuVista+, don't forget to feed the composite video input with a blackburst signal.

This is an exhaustive look at setting levels, but they must be right. This is the biggest stumbling block for many editors who have never been required before to care. Look at levels at every stage of the finishing process. What does that imported graphic really look like? I'll bet the print graphics designer didn't take into account broadcast chroma levels when they designed the logo. Train your eye to see color on your carefully adjusted monitor and scan for reds and yellows that look too hot and double-check them through the scopes. You will learn what to look for and you will feel much more comfortable with the whole process.

## AUDIO LEVELS

Input and output audio levels must be analyzed before you do anything with audio. Generate a 1 kHz audio tone at 0 dB analog and input it to the Avid. There really is no industry-wide agreement on reference level for the digital level for audio reference tone and it may vary from -20 dB to -12 dB. If you don't have your own specific requirement, or your audio engineers don't have a preference, then go with the Avid default of -14 dB. It is close to the middle of the range and, although may not give you as much headroom before you distort, -14 dB is fine for a mix of different types of audio, both conversations and explosions. The digital reference level is adjustable in the Audio Tool under the right-hand pull-down menu (PH), Set Reference Level, and the scale on the Audio Tool will change.

Playback the videotape with the 1 kHz tone at the proper level by adjusting the audio playback controls. If you do not have these on your deck, you need to double-check that anything recorded at the reference level of 0 dB analog also plays back at that level. Otherwise, your deck needs adjusting first. Then you must decide if you are going to use a mixer during the input stage. If you have lots of different audio sources like DAT, cassette, CD, or microphone, it might be better to run everything through the mixer. This also gives you a place to equalize or affect the audio one shot or one tape at a time. But now you must calibrate the levels through the mixer. Make sure that the 0 dB tone from tape is at 0 dB in the mixer when both the individual faders and the master faders are in the detent or notched position. Many mixers have trim pots for each input, little knobs in the back, where each source can be adjusted individually as they come into the mixer. Be sure to pan odd numbered channels to the left and even numbered channels to the right.

Finally, look at the Audio Tool. Is it at 0 dB too? If not, you have two choices depending on the hardware of your system. Whatever you do, do not touch the slider in the Audio Tool with the Speaker icon. This adjusts the monitoring level, not the input level. This should always, always be at 0 unless you are specifically adjusting audio output that cannot be raised by any other means.

If you have a system that only allows two channels of input at a time, you either have the Audio Media II (NuBus) or Audio Media III (PCI) cards. These have an adjustment for input already in the Audio Tool that has a Microphone icon. Bring the signal to the 0 dB analog level in the Audio Tool with this input slider.

If you can input four channels of audio at the same time you have the SA4 (NuBus) or Amadeus (PCI) audio cards (8-channel input is possible with the 8×8 from Digidesign with MC Version 7.0 and PCI). This is a little trickier since it involves adjusting the levels by tweaking the Digidesign Pro Tools 4×4 that came with this system. These input and output levels should be set perfectly at the factory and really shouldn't have to be adjusted, but if you are setting it up without a mixer, there may be some changes necessary. The benefit to adjusting the Pro Tools 4×4 is that once the hardware levels are set, no one should touch them again. The Pro Tools hardware requires a small screwdriver, sometimes called a *tweaker* or *greenie* because it is used only for making these little tweaks and is usually green plastic. The larger, Avid screwdrivers used for assembling the system are just a little too big and may strip the screw adjustment if you are not extremely careful. Carefully tweak the input levels until there are no yellow LEDs lit in the Audio Tool when inputting the reference tone. For a quick look at where the LEDs should be, call up the reference tone generator in the Audio Tool and observe where the perfect tone should be. Then, with the levels adjusted perfectly, digitize a minute or two of this tone into the system.

Playback the digitized tone from the Avid and see if it plays exactly the same way it was digitized. It should. Now stop that playback and call up the reference tone again from the Audio Tool. Where does your audio go out of the Avid? If it goes through any distribution amplifiers, these amplifiers should be checked out for unity. Then, does the signal go to the mixer again or directly to the deck? My vote is for it to go to the deck, but many people like to ride the levels one more time as the show goes to tape by putting it back through the mixer. This is fine, but it adds another level of complexity. It also means that there will never be two identical digital cuts of the same sequence unless the levels in the mixer are not touched. My approach requires getting it right in the Avid and then outputting as simply as possible, direct to the deck, to avoid those inevitable 2 AM mistakes.

If the output of the tone goes through the mixer, put all the mixer faders to their notched position again and look at the level. Is it 0 dB? If not, adjust the trim pots for MC/AX inputs to the mixer. If you are not going through the mixer, then look directly at the deck. Set the deck record levels so that they record the tone at a perfect 0 dB (or the proper digital equivalent if you have a digital deck). Here, if you have the Pro Tools 4×4 and are going directly to the deck, you can cheat a little if you want. If you have a separate tone generator, you can input directly to the deck and set the record level. Then see if the level coming out of the Pro Tools 4×4 matches this. If not, you may want to adjust the output levels of the Pro Tools. Alternately, if you trust the deck preset recording level, adjust the output levels of the Pro Tools to be at 0 dB when the input setting of the deck is in preset.

There is a twist if you are using the Audio Media II or III card which bears repeating. These two cards work with unbalanced audio and so must use cables that go from XLR to RCA connectors. They must also work with the input at -10 dB lower than the output. This is taken care of if you use the Avid cables that shipped with the system. If you must use other cables for some reason make sure they have a -10 dB attenuator or pad on the input to the audio card or you will never get the input and output levels to match.

Let's go back to our digitized tone already on the Avid and play back tracks 1 and 2 only. Look at the levels on the record deck and note where they are; they should be 0 dB. Go to the Audio Mix window and pan both tracks 1 and 2 to the right. How much does the level rise? About 6 dB. Now turn off the monitor for channel 2. A drop of 6 dB on the record deck levels. Keep this in mind as you mix your tracks. As you add audio tracks to the mix and then mix them down; the overall level gets louder. If the stereo audio comes from a digital source like a CD and the mono sound comes from an analog source, you may have to drop the digital source level because it will have more headroom — the loud parts get louder than an analog source is capable of reproducing. If you have a sound compressor, you may want to compress the CD music so it is more even in level and does not get too soft or too loud in comparison to other analog sources. Some CD library music is already compressed for this reason.

## SIMPLE REDIGITIZING

When it is finally time to take the final cut and redigitize at the finishing AVR, you have two choices of action. Take all of the low-resolution media offline (by either deleting or moving) and redigitize the sequence using batch digitize. Highlight the sequence, change the AVR, and choose Batch Digitize from the Bin menu. First, the system asks if you want to batch redigitize everything or only the media "which is unavailable." Because you have taken the media offline, all of the media is unavailable. The system then can keep track of what has been redigitized so when you stop and start, it knows what has already been digitized (that media will be "available") and the system skips those clips.

The system asks you for the tapes it needs and takes every shot it needs from that tape. The beauty is that it only digitizes as much of the shot as it needs to play the sequence with a little extra thrown in as *handles*. The handles give you enough media to make minor changes like the length of a dissolve or allow you to slip the shot a few frames to avoid a dropout. Do not underestimate the importance of handles, but do not forget that you are using up disk space very quickly at the high AVR. At the end of the batch digitizing process, you will have a bin full of new master clips that have the original clip name plus a ".new" and a number at the end. This is your new high-resolution media.

Batch digitizing the sequence is simple, but it doesn't give you all that much control. If you have a very complex or long sequence, you are at the mercy of the system in terms of stopping or pausing or even trying to anticipate what the next needed tape will be. If you leave all of the low-resolution media online when you start digitizing, because you will continue to cut low-resolution versions after the online, then you cannot take advantage of the "only digitize media which is unavailable." Since all shots will be available and the system does not distinguish between AVRs, when you stop digitizing and want to pick it up later, it is not so easy. You have to skip over every shot and every tape already digitized or every tape not present, one shot or one tape at a time. Especially if you have skipped tapes on purpose, maybe because they are not ready, you have to manually skip over them with "skip this tape" each time you stop and the system asks for it again. With several hundred tapes this can get old fast.

## DECOMPOSE

Decompose is the feature you can use with Media Composer to specifically address the need for redigitizing sequences. It is designed to give control back to you by breaking the sequence into new master clips before digitizing. The basic logic of the system tries to protect you from digitizing over material that is already there. This is why, in the first scenario, you should take media offline before you batch digitize. When you decompose, the low-resolution media is taken offline automatically for that sequence when the new master clips are made.

Duplicate the sequence and put it into its own bin. Highlight the sequence and choose Decompose under the Clip menu. When you decompose, you get a warning that is a little intimidating about how this function will take all your media offline. Always duplicate your sequence before redigitizing; this message means that, for this sequence only, it will break the links to the media already on the drives. In one step, it creates the new master clip with the ".new" on the end and, since these new master clips have never been digitized before, they link to nothing on the media drive and so all media is considered "currently unavailable." There is no need to take any media offline before batch digitizing.

Now you can arrange the clips in any order you like. You can easily skip over tapes that are not ready yet and sort or sift the new master clips in the bin. When you finish batch digitizing you have all the high-resolution media you need to play the sequence plus handles for minor changes. We will come back to the Decompose option later in this chapter because it is so important as the first stage of some sophisticated variations on this simple redigitizing process.

If you have titles, graphics, or animation and you cannot create or import them at the finishing resolution the first time they are used, then you must reimport them at the higher resolution. All titles must be recreated, an automatic process that leaves them intact, but changes the AVR. All imported graphics with alpha channels must be recut into place by eye. The graphics and animations that have no alpha channels can be automatically relinked by a method I will describe later. All effects must be re-rendered; then everything is digital cut to a tape prepared with timecode.

## MORE COMPLICATED DIGITIZING SCENARIOS

### Drive Space and Offline Media

Your strategy for redigitizing is going to first require a decision about drive space. Do you have enough space on your media drives to keep all of your low-resolution media? This is a critical decision since you may need to recut sections and make multiple versions after the final sequence is finished. Maybe your producer will want to make changes after viewing the final high-resolution master and you will have to go back to the source material to find new shots. Deleting your low-resolution media is a big step toward the point of no (quick) return and should be done only after serious consideration. It also significantly impacts the way that you manage your media during the redigitizing. Let's look at several scenarios depending on whether you keep or discard media while streamlining and minimizing the redigitizing process.

### *Splitting the Sequence*

In the first scenario you have enough space for all the media to be on the drives at the same time. The decision is also to keep all of the audio exactly as it is right now. There has either been lots of sound mixing or the mix will be done by someone else. Either way, there is really no reason to redigitize audio unless it is distorted or is being replaced with material that is much better quality, say the DAT original instead of a U-Matic dub. If the audio needs to be redigitized, the mix has to be tweaked because the levels during the redigitizing may be just slightly off what they were originally. You can save significant amounts of time and money if you can do anything to keep from redigitizing and remixing your sound. Think seriously about changing your offline methods so that the audio does not need to be redigitized during the finishing stage.

This strategy entails stripping the video away from the audio when starting the redigitizing process so that you are only working with video clips. There are several easy ways to do this, but let's use the easiest first. Duplicate the finished sequence and put it into its own Redigitizing bin. Load the sequence into the Record side of the Source/Record monitor, highlight just the audio tracks, and hit Delete and OK. You are left with only the video tracks. Now highlight the same sequence in the bin and choose Decompose from the Clip Menu. There will be a checkbox that says, "Decompose only those items for which media is unavailable." What this really means is, should this function create new master

**Figure 8.5** Decompose Dialog Box

clips for everything in the sequence or just the media that is missing or offline? Since everything is available or on the media drives and linked to the sequence, then leaving this box checked results in nothing happening. This can be used effectively in other strategies, but not now. This is not the desired outcome, so uncheck this box and, whether the media is available or not, the Decompose function creates a new master clip for every shot in the sequence and adds a ".new" to the end to make sure there is no confusion with the original (longer) master clip.

### Using Media Mover to Take Video Offline

There is another way to take video offline that does not require splitting the original sequence into two separate sequences of audio and video. Chapter 4 discussed a third-party program called Media Mover. This program can search all your media disks and find all the media for all the projects used in this sequence. Media Mover versions 2.0 and later have the capability to move just the audio or the video out of the MediaFiles folder and into a separate folder with the project name. This effectively makes the media offline or unavailable and, if you do this to the video only, when you use the Decompose function, you can leave the "only decompose clips for which the media is unavailable" checkbox checked. Media Mover looks for the parts of the sequence that are offline, the video that you just moved, and makes ".new" master clips for the video only. Either way, splitting the video and audio tracks or moving the video out of the MediaFiles folder, you end up creating new, offline, video-only master clips.

## Handles and New Master Clips

Decompose may not actually create a new master clip for each shot in the sequence. One exception is imported graphics and animations. These are not decomposed because they can not be batch digitized. Another reason you may not get a new master clip for every shot is the second part of the Decompose dialog: the handles.

The handle length defaults to 60 frames and is easy to change. If you shorten it and you have hundreds of cuts in your sequence, you save yourself considerable disk space. 100 clips times 120 frames per clip for handles adds up to an extra six-and-a-half minutes of footage in NTSC. If your one-hour show has 600 to 800 edits, this is a serious amount of extra storage space at high resolution. If you make the handles longer, then two shots in the sequence that were right next to each other on the source tape may actually overlap with timecodes. The Avid determines if there is an overlap and combines the two (or more) shots into one master clip and so reduces duplicated frames. Combining multiple shots intelligently into a single master clip reduces the amount of master clips and speeds up the redigitizing process by eliminating the amount of time needed to stop, search, and preroll for each shot individually. Extrapolated over hundreds of shots, this can really add up to a time savings.

## Automatically Offline

When you decompose, the sequence breaks its links with the low-resolution material and links to the new offline master clips. The ".new" master clips are offline and they are video only. You could sort and sift, stop and restart the batch digitizing to a higher resolution and have complete control over the process. But there is one drawback: What if you are asked to continue to edit with these new AVR 77 video-only clips? Suppose you are asked to make a new version of the sequence? Where are you going to get the sync sound? From the original audio? And then sync every shot by matching timecode? All of this seems like a huge amount of work. Working with video only can backfire and force you to lose the savings created by leaving the audio alone. If you are totally sure that you will not be asked to continue editing at all with this material, then go right ahead and start batch digitizing. But if you know your client too well, there may be an extra step.

You can modify the video-only master clips and add audio tracks to them and be sure to always have sync sound. You will not use this audio for the final mix! The levels do not have to be recreated perfectly, the new audio is for reference only, and you will not accidentally cause the sequence to link to this new audio. It is only used in case you are asked to go back to the high-resolution shot and play the whole thing through. If you match frame on the video track in the final sequence, the system calls up a shot with sync audio. If you match frame on an audio track in the sequence, you call up the original master clip with the original audio. You really have the best of both worlds with one minor drawback: Redigitizing the audio takes up more drive space. But audio really takes up very little space compared to video and if your sequence is really not that long, the difference in storage in minimal.

Take the extra amount of audio into consideration along with the extra amount for handles when redigitizing. One way to figure out how much space you need is to select all the decomposed master clips in the Decompose bin and hit Command-I (Get Info). In versions 6.5/1.5 and later, this calls up the Console and gives you the total length of all the clips in the bin. In earlier versions of the software, you need to open the Console first before using the Get Info.

Continue with the batch redigitizing and keep an eye on your levels. Some people actually make the preroll time a little longer during this stage so they can quickly correct any little level problem they see before the shot starts to digitize. This is why you take the deck adjustment out of preset. Many times the shot requires a tiny raising or lowering of the video gain to even things out. Remember that the settings created during the offline are automatically recalled during this batch redigitizing process. You may not want the tape levels setting to be recalled! Perhaps someone set the tape up incorrectly or you are using the original Betacam SP reel instead of the S-VHS dub and the two tapes are quite different looking because of a bad dub. You have to create a new setting with the same tape name to replace the existing one or go to the Project Window and delete the tape setting completely.

The tapes are requested by the Avid one by one. They are used from the earliest shot on the tape. Then they fast forward through to the end of the reel and take all the shots that are needed. Each tape is only used once. You can speed up this process by either rewinding all the tapes beforehand or actually cueing up to the first shot that is needed on that tape. This is easy to do if you have an extra deck or two and a printout of the decomposed bin.

### *Putting the Sequence Back Together*

Once the high-resolution video and reference audio are digitized, put the video-only sequence in the Source side of the Source/Record monitor and load a copy of the original sequence into the Record side. Cut the video only from the high-resolution sequence over the low-resolution video in the original sequence. The new video should match perfectly and have the same starting timecodes. If you hit the Home button and mark in on both sequences, the new high-resolution sequence will be in perfect sync with the original sequence. Even the sync breaks work because they are looking for tape name and timecode. So if you continue to tweak the sequence and you knock yourself out of sync, you are still given the white numbers in the timeline. This may be the last step for a simple sequence.

## The Low Disk Space Variation

When you have very limited disk space then all low-resolution media must be deleted and the audio must be reduced to the bare minimum to free up space for the AVR 77 material.

Here, you can use the Media Tool to delete large amounts of low-resolution video all at once. Choose to show all the media files and precomputes for this project. Select all the media in the Media Tool and hit Delete. A dialog box appears and asks you if you want to delete audio or video. Uncheck the audio choices and delete all the video only. Scan through the sequence to make sure all of the video is gone. If there are any shots left, then they were part of another project. You must decide whether they can be deleted or not. To be completely sure this method will work, you should either delete these shots or figure out how to take them offline. You can do this by dismounting the media drive they are on or by moving the media from those projects out of the MediaFiles folder using Media Mover.

This is a variation of the first method and is similar except that you have deleted the video media instead of just moving it out of the way. At this stage, if you are really tight for space you can consolidate the sequence. This takes the audio media and copies only the parts you need. Be sure to uncheck "Skip if media is already on the drive" because you are not just moving the media, you are shortening the original master clips down to only the parts you need. You can check "Delete after consolidating" if you are sure you have enough space to consolidate the entire sequence on a single drive. Otherwise, split the sequence into pieces and consolidate each piece to a separate drive. Versions 7.0/2.0 allows you to consolidate to a list of selected drives. Go back after the consolidation and delete the original audio media by hand; this method is slower but you have more control.

Now you have a sequence that is linked to the consolidated audio and you can decompose the video only because, having been deleted, it is offline and unavailable. If you have the space to redigitize more audio, you can modify these video-only clips as mentioned before to add audio tracks or just go ahead and start to batch redigitize.

Here is a scenario that Basil Pappas solved with the best method I have seen. It involves mixed AVR 12 and AVR 77 material when you want to keep all the AVR 12 material before redigitizing.

## Mixing AVR 12 and AVR 77

If you don't need the AVR 12 material you can sift for it in the Media Tool and delete, but now you need another way to separate the AVR 12 from the AVR 77 because you want to keep all of your media available. Strip the audio tracks from a copy of the sequence and make it a video-only sequence as in the previous technique. Put the new sequence into a new sequence bin. Open the Media Tool and show master clips from all projects. Make sure nothing is highlighted or selected in the Media Tool. Highlight the new video-only sequence and, in the new sequence bin, use the Bin menu choice, Find Media Relatives. This highlights all the clips in the Media Tool that are used in the sequence.

Drag all the highlighted clips from the Media Tool into the new bin. Sift the clips in the new bin so that only the AVR 12 clips are showing and select them all (Command-A). Now you are going to unlink these clips from their media and purposefully make them go offline. If you hold down the Control and Shift keys at the same time and look under the Clip menu, you see that Relink becomes Unlink. Choose this and all the AVR 12 material unlinks and becomes unavailable.

Move the video-only sequence into yet another new bin and decompose. Leave "Decompose only the clips for which media is unavailable" checked and the sequence will decompose only the offline AVR 12 video clips. Decompose creates new master clips for all the original AVR 12 clips and it leaves everything else in the sequence alone. Now go back to the bin with all the unlinked AVR12 clips, select all, and relink them. Be sure to check "Relink masterclips" in the relink dialog. All the AVR 12 clips should go back to normal. Begin digitizing the decomposed, video-only master clips. If you like, you can also take the extra step and modify the video only master clips and add audio.

### Decompose without Timecode

What if there are shots in the original sequence that have no timecode? If they were decomposed too and unlinked from the sequence because of decompose, how do you get them back? This scenario is possible if you were working at AVR 75 or 77 for the material that had no timecode and were mixing this with AVR 12 for the material that you were planning on redigitizing. You don't want to redigitize the high AVR material but it is decomposed anyway (unless you used the previous method). When the batch digitizing process is done there are gaps in the sequence where the non-timecoded material should be. You can check to see if you have any gaps in your finished sequence by putting a copy of the redigitized sequence into another bin and decomposing the sequence again. This time leave the box checked for "Only decompose the clips for which media is currently unavailable." All the media should be available now except those shots not redigitized. If you missed a shot for some reason, it is a new decomposed masterclip. This is a simple way to pinpoint what is missing.

When non-timecoded material was originally digitized, it was given a time-of-day timecode that can not be recreated. This timecode is made up by the system and assigned during the digitizing and is not really on the tapes. This fake timecode is enough so that the system thinks this material is different from a graphic, video mixdown, or another kind of imported media that has no timecode at all. Imported media is not decomposed, but the digitized, non-timecoded material is.

Consider this: The sequence can always be relinked to the original media even after the decomposing. If the redigitizing of timecoded sources is complete and you relink the sequence, the sequence stays linked to the new, high-resolution images and looks to fill the gaps left over from these last, non-timecoded shots. When the sequence tries to link, it is looking for timecode and tape name, finds the original time-of-day clips, and reconnects them with the finished high-resolution sequence.

The goal of all these methods is to reduce the amount of redigitizing to a minimum. These strategies require selectively removing only what needs to be replaced. Many people are content to spend the extra money on drives and always work at high resolution. This is a perfectly fine way to work although the Macintosh may not react quite so quickly to commands that have to do with recalling and playing back all that high-resolution media. The computer handles the smaller, low-resolution files more quickly and overall performance, like rendering, is better at the lower resolutions. As the drives and the computers get faster, this will be less of an issue, but for now it is something you might take into consideration as your project gets longer.

## Recreating Titles

If the titles created from the Title Tool and used in the final sequence were first created at the low resolution then they must be recreated at the finishing stage. This is an automatic process that takes little time and replaces the old titles with new, high AVR versions. There are two parts to every title, the type information about drop shadows, borders, and alpha channels and the title media created at a specific AVR when you leave the Title Tool. The type information must be used again when you recreate the title media and new media is created based on what is set in the Compression Tool. So set the compression tool to the highest AVR possible, mark inpoints and outpoints at the beginning and end of the sequence, highlight all the video tracks, and choose Recreate Title Media from the Clip menu. The MC/AX looks through the sequences, finds all of the titles, generates new media, and links it to the sequence.

In the process of recreating titles there is a warning that says, "This process will remove any nested effects." This refers only to nested effects within the titles. Nesting video in a title is a very cool effect and easy to use. You fill a title with moving video by stepping into the title and editing new video over the unlocked video track (V2) and maybe you add some other effects inside. The problem when you are recreating the titles is that the MC/AX is going back to the original title information. The original type information knows nothing about what you did to the title after it was created. In versions 7.0/2.0 this will make a title lose its uncompressed status.

You can preserve the work you did inside the title by subclipping the nested effects, dragging them to a bin, and then placing them back in the title after it has been recreated. The only way to subclip material from the inside of a nest in MC version 6.5 is to use the clipboard button. Mark-clip the entire material inside the nest and hold down the Option key while clicking on the clipboard button. This puts the material automatically into the source monitor. (Since the only time I use Clipboard is as Option-Clipboard, I have mapped the option button—a little dot in the command palette with the letters OP—onto my Clipboard button.) You can then use standard subclip techniques to drag the nested material to a bin.

In versions 7.0/2.0 you have the option to use uncompressed titles, rolls, and crawls. You can't promote these uncompressed titles to 3-D, but you can play them in real time while there is another video track below with unrendered real-time video effects. Rolls and crawls are as easy as hitting the R or C button in the Title Tool and typing away.

## Replacing Graphics

The final step for "rezzing up" is to bring all the graphics up to the finished quality AVR. If your imported graphics have alpha channels then they are considered "effects with source" and are treated differently than straight graphics. Unfortunately, at this time, the only way to change the AVR on these alpha channel graphics is to reimport them and cut them back into the sequence one by one. This can be done a little more quickly using the Replace function in Media Composer. The Replace function is the blue arrow available under the fast menu in the center of the Source/Record window or in the Command Palette. Use Replace instead of Overwrite. Replace works well because it requires no mark inpoints or outpoints. In fact, any existing points should be deleted. Put the blue position bar close to the beginning of the new, high-resolution graphic in the Source window, position the blue position bar in the timeline over the graphic, and choose Replace.

There is another method for fast replacement of material. This method edits directly from the bin to the timeline. Use the mark clip button to mark the low-resolution graphic in the timeline. Choose the red segment mode arrow and drag the graphic from the bin to the timeline. Hold down the Command key and the graphic snaps to the inpoints and outpoints in the timeline.

If you created keyframed moves for these graphics (including fading up and down), you want to preserve that work for the new high AVR replacements. To save the keyframes for a regular graphic, drag the effect icon in the upper left-hand corner of the effect mode palette to the bin. This saves the keyframes without the source material when there is no alpha channel. To save the keyframes from an imported graphic with an alpha channel, Option-drag the effect icon from the effect mode to a bin.

## AUTOMATICALLY RELINKING GRAPHICS WITHOUT ALPHA CHANNELS

I have developed a way to automatically replace graphics or animations that do not have alpha channels. If there are only a few graphics to replace, the two methods just described are probably the best and the simplest. But if you have 20, 50, or 100 graphics you need to do a little more work to set up a process that forces the sequence to break the links with the old graphics and relink to the new ones.

The trick here is that to properly relink to a sequence, you need timecode and tape name to match between the source in the bin and the clip in the sequence. When you import a graphic the timecode is always the same: 1:00:00:00. But what is the tape name? There isn't one and that's the problem. You can give the imported graphic a tape name using the modify function! If you give all the graphics the same tape name, that would be simple. Just select them all and modify them all at the same time. Unfortunately, they would all look exactly the same to the sequence, same timecode and same tape name. This would link all the graphics in the sequence to the first graphic modified—all the old graphics would be replaced by one new graphic! So you must give each graphic a unique tape name, then break the link between the old graphic and the sequence and force a relink to the new high-resolution graphic. If you used many PICT files or other graphic formats and have combined them with effects, this method saves you lots of time compared to replacing them by hand.

Determine the graphic files needed in the finished sequence and isolate them in a bin. This can be done by opening all the bins with graphics or the Media Tool, selecting the sequence, and use Find Media Relatives in the sequence bin pull-down menu. The graphics used are highlighted and can be dragged into a new Used Graphics bin to isolate them.

Click on the name of the first graphic (not the icon) and copy the name into the Macintosh clipboard (Command-C). Click on the icon of the graphic to highlight it and go to Clip menu and choose Modify. In the Modify dialog, choose Source and paste the name of the clip as the new source. You are limited to 32 characters and no punctuation, but it's a start. The idea is to make the source name simple and obvious. Press OK at all the messages about how terrible this is and you will probably die penniless, etc. Do this for all the graphics you need to relink. Actually, modifying master clips is pretty serious but in this case, you need to modify a lot of individual master clips as fast as possible. You may consider programming a macro to just click OK on all those messages.

Import the necessary graphics one more time at the higher AVR into another Reimported Graphics bin. Modify the sources of the new graphics and this time, when you go to Modify source, you see the choices for source names already in the Choose Other Tapename window. Double-click on the obvious names that match the name of the graphic and you have saved yourself all that cutting and pasting.

There is a new twist to this method in later versions of the software since you can specify the length of the graphic during the import process. You must import the low-resolution and the high-resolution graphic files at the same length. If you always leave it at the default length of 10 seconds, you can't mess this up.

Now just delete all the original low AVR graphic media in the Media Tool and relink the sequence. Make sure none of the boxes are checked in the Relink dialog. All your graphics in the sequence are relinked to the higher AVR media and all of your effects that involved the imported graphics are intact (although unrendered).

The more complex procedure is to leave all the low-resolution graphics online and force the sequence to relink to the high-resolution graphics. Drag the sequence into the Reimported Graphics bin with just the high AVR imported graphics and select all. Relink the sequence with "Relink to Selected." This allows you to keep all the low AVR graphics on the drives linked to other versions of the sequence.

## Rendering

After all the graphics and animation have been reimported and linked to the sequence, it is finally time to render. Most people don't realize that they only need to render the top track of any effect. As long as the top track covers the entire effect, you do not need to render all the tracks below. This saves a lot of time, but people insist on rendering everything anyway. It makes them more sure that they will not get a video underrun in the middle of a complex effect sequence and, if you have the time, this is foolproof. Rendering everything may also speed up access time when you are moving between layers because the system never has to render on the fly anymore; it can just call up the rendered precompute. But if you are making lots of changes or are in a hurry to just get something out the door, then all this extra rendering can be skipped.

## Timecode on the Master Tape

When you are ready for a digital cut, you need to consider how the timecode on the master tape has been set. The default timecode for sequences is 1:00:00:00 so you should either black the tapes so that you have enough space for bars and tone and start the show at this sequence time or change the start time on the sequence. If you start blacking your tapes at 58:00:00 or 58:30:00, then you have enough room for a minute of color bars and tone starting at 58:45:00 and then 15 seconds of black or countdown. (Actually, you have 13 seconds since there should never be anything in the last two seconds before the program starts.) This uses the sequence time option during the digital cut. This method is preferable because it means there is a permanent relationship between the master tape and the sequence that can be relied upon even years later if something has to change or you want to edit into the master to make a small change without redigitizing the entire sequence.

For editors who are working with tapes blacked and timecoded with no regard for the Avid settings, you will have to change the sequence time of all the sequences to match the tape. Many editors leave their Betacam decks in the preset timecode mode and whenever they do a crash record to start blacking the tape, the timecode starts at 00:00:00:00. By the time the bars, tone, and countdown are finished, the starting time of the sequence is around 00:01:30:00 or 00:02:00:00. You can change the starting time of all future sequences by going to the General settings and changing the time there. That General setting can also be put into the Site Settings. General settings are also the place where you change the default sequence time from non-drop frame to drop frame. Highlight the number already in the setting and type a semi-colon. The type of timecode on the master tape, drop or non-drop, must match the type of timecode of the sequence or you will get an error.

If you are putting multiple sequences on the same master tape, you may not want to take the time to change the start timecode on each sequence. In that case, there are the other two choices in versions 7.0/2.0 Digital Cut window: Tape Time and Mark In. Tape Time is wherever the tape happens to be cued up at the moment; the tape backs up and prerolls to insert edit at that point. Mark In (in earlier versions, Custom Time) allows you to type timecodes in the deck controller at, say, regular intervals of two minutes. If you are using Custom Time in versions prior to 7.0/2.0, the type of timecode in the Custom Time window must match the type on the master tape. You must type the semi-colon when you are inputting the Custom Time number if you have a drop frame master, but once you make the change, it stays that way until you quit the application.

## ASSEMBLE EDITS

Assemble editing has long been a feature request and the addition of this function has saved many a session from stopping just to black a tape. Using an assemble edit also saves many hours of tape deck record head wear and tear.

Why would you use assemble editing? If insert editing was the only choice, you could not perform a Digital Cut until your one-hour show had a prepared black and encoded one-hour tape (timecode laid down over the entire tape while recording black). Maybe the plans change at the last minute and now there are two versions of the program, one with titles and one without. Do you have two blacked tapes? If not, don't black another tape before you begin, perform an assemble edit for the digital cut.

An assemble edit lays down new timecode during the edit. That timecode is determined by the settings on the record deck. There must be enough timecode on the tape for the preroll; a minimum of three to five seconds that you can set in the Deck settings. But since the worst quality of any videotape is at the very beginning and the very end, give yourself a healthy thirty seconds if you can afford it. All professional videotapes actually contain well over two minutes more than the length on the outside box would imply. After blacking just that thirty seconds at the beginning of the tape, you can perform an assemble edit and the digital cut lays down the new timecode during the playback of the program.

If you set the deck correctly to start the tape blacking at 58:00:00 or 58:30:00, then during the preroll for the assemble edit, the deck reads that code and regenerates it. The deck setting to start blacking the tape with correct timecode should be preset but then the switch must be changed to Regen before the assemble edit is performed. If you do not change the switch from Preset to Regen before the assemble edit digital cut, then you create timecode 00:00:00:00 at the edit point. Try to do another edit at that point and see what happens when you search backward from 0. The deck thinks it must rewind the 23 hours and two minutes to find the number for the inpoint, 58:00:00.

Where is the choice for assemble? If it is not in your Digital Cut window then it must be turned on. Go to the Deck Preferences setting and notice the checkbox for Enable assemble editing. Now you have a pull-down menu in the Digital Cut window with two choices, Insert and Assemble.

What happens if you do an assemble edit instead of an insert edit in the middle of the program? This is the most serious mistake you can make with a master tape and it is the reason why the assemble edit function is not only just a little hidden, it is displayed in red. When the assemble edit is over, there are several seconds afterwards where the control track of the tape is unstable as the recording comes to a stop. This can cause any image on that part of the tape to be unstable and unusable and this cannot be fixed except by continuing the assemble edit to the end of the entire sequence. You can not perform a clean insert edit over a point in the tape where the control track is unstable, like at the end of an assemble edit. You will always have a glitch. You should only use the assemble edit mode if you are planning to lay the sequence down all the way to the end and there is nothing else on the rest of the tape.

## OUTPUTTING TO NON-TIMECODED TAPE

If you are outputting to a non-timecoded master tape, you cannot control the record deck from the computer. This method is strictly for a quick crash record or going to VHS for a preview copy. Use the countdown built in to the software to make the start of the presentation look a little cleaner and more professional. This countdown is purely for the purpose of letting the tape come up to speed; it is not frame accurate and is slightly different each time you use it. If you want a frame accurate countdown, make one and edit it onto the front of the sequence. Spend a few minutes and change the PICT file for this countdown if you have a company logo. Any correctly sized PICT file for your video board and format will do and it adds a nice touch.

## COLOR BARS AND TONE ON THE MASTER

Where should you get the color bars and tone to put on the head of your master tape? One thing is for sure: bars and tone should be representative of the levels in the show that follows! Just sticking a pretty picture with lots of colors at the beginning of your show doesn't make the show look good. If you set a deck up by using these bars, will the show be legal, broadcastable, and listenable?

Here are a few choices that allow you to decide what works best for you. You can output both a perfect color bar signal and perfect 1 kHz tone directly from the computer as discussed before when setting all the levels in the edit suite. If you perform an insert edit using the edit controls on the front of your deck, you can lay both bars and tone to the head of the tape simultaneously. If you are confident that your video and audio that follow are just as perfect, this works out fine, with one exception. You want to make sure that there is no slight jump in black levels between the black on the master tape and the beginning of the fade-up at the start of the sequence. The little jump in black levels is visible to some people when they have the black levels turned up a little too high on their monitors at home. You could add two seconds of black filler to the start of your sequence and then change the start time of the sequence to 59:58:00. You can play back black filler from the Avid and use the front panel of the editing deck to insert edit onto the tape after the end of the color bars. Or you could splice all the bars, tone, and countdown to the beginning of the sequence and change the start time to 58:30:00.

## IMPORTING COLOR BARS

The best way to get absolutely, indisputable color bars and tone into the MC/AX is to import them. There are two sets of color bars available to import that come with the system. Both have serious implications and should be used only under the right circumstances. The first graphic that can be imported is the one recommended by Avid. It comes in the Goodies folder on the disk with the rest of the software and utilities. This graphic is called Color Bars for Import and is designed to be imported with RGB levels. Using RGB levels means that this graphic is for versions of the software and hardware that do not give you the choice of CCIR values during import. There is nothing wrong with this graphic except that it is designed to be used in NuVista+ systems that cannot use any black level below 7.5 IRE. As discussed earlier, the split field SMPTE color bars have a pluge that descends to 3.5 IRE. This lowest pluge is not included in these bars. This doesn't affect the usefulness for setting up a signal, it means you have one less tool for adjusting the brightness of a monitor. If you have a NuVista+ system or your software version can not import at CCIR, this is the only graphic you should import for correct video levels.

There is another set of color bars designed to be used only with the Output Tool. If you call color bars up as the choice for a reference signal directly out of the Output Tool, you can adjust your output settings to be perfect. This color bars PICT file is created at CCIR values and is meant for direct output only, not importing. If you import these files, you need to make sure that the import settings are set to CCIR values and forced to fit. Otherwise, you have bars that are dangerously wrong. The white levels are around 80 IRE and the blacks are around 14 IRE. I have heard too many stories of the dubs looking wrong or the show being rejected and the editor always claims, "These were the bars from the Avid!" Yes, but they were used wrong. Let me repeat: If you import the color bars from the Settings folder, you must import with CCIR values and use the Force to Fit option.

If you are confused or suspicious about what color bar levels you have, check the bars at the head of the sequence, either through the internal Avid waveform monitor or through an external waveform and vectorscope. If the color bars generated by the Output Tool look perfect, but the color bars at the head of your sequence do not, then the bars in the sequence must be replaced.

Audio tone must also represent the level of the audio to follow. Versions 7.0/2.0 gives you the ability to create a perfect tone media file in the audio tool. This can then be edited onto the front of the sequence with the imported color bars. All of this preparation can be done as a last step before performing the digital cut. Personally, I hate editing with a sequence that has all of this reference information at the head so I put it off until the very last stage. When I use the Home button, I want the sequence to go to the first frame of the show, not color bars.

There is another school of thought that says that the color bars must be digitized through the video board if they are going to match the rest of video that has also been digitized through the video board. It is much more important that the levels all be the same throughout the sequence and run through the same scopes. This argument also does not take into consideration that a substantial portion of your sequence may have been imported and not digitized at all! But just for convenience, you may find that carefully digitizing color bars, tone, your own special slate, and countdown and black is just easier for you to edit onto the head of your sequence. If the levels are not representative of the video that follows, this method is not any more accurate than the other methods.

## THE FINAL CUT

When it is time to actually perform the digital cut and everything else has been prepared, there is one more setting to check. Note when you call up the Digital Cut window which tracks are highlighted. You can patch the audio or video layers from the Avid to the tracks on the tape, making it easier to customize the master tape without making multiple sequences. Maybe you are going to add titles in a last digital pass in a linear suite because you have a specific look that can only be created by a Chyron character generator. You want to lay off a copy of the show without the video track that contains the titles. If it is an assistant who will perform this function, they must be quite clear on the implications of this setting.

What if you are viewing the digital cut one last time as it goes to tape and discover a minor mistake? You can perform an insert edit in the middle of the sequence without laying down the entire sequence again. Make sure you have the type of edit set to insert in the Digital Cut window first. Then mark inpoints and outpoints in the sequence at existing cut points before and after the change. Then uncheck the Entire sequence checkbox in the Digital Cut window and highlight only the tracks that need to be changed.

If the change lengthens or shortens the sequence, be aware that there will be some more work to do at the end. If the sequence is now shorter after the change than it was the first time you recorded it to this tape, then you must insert black and silence onto the tape after the sequence is over. There is almost nothing more embarrassing than fading to black and then coming back for a little reprise of the last few seconds! You can add black to the end of the sequence, but it is not as easy as you might imagine. You cannot just add filler to the end of a sequence in the Avid. There are a few workarounds to choose from if this is desired, but I prefer the simplest method. Use the edit controls on the front of the master deck and do a short insert edit while playing back filler from the MC/AX. This will not work if you don't have direct access to the deck or your deck does not have front panel edit controls so you may be forced to use one of the following methods.

You can import a black graphic or digitize black and edit it onto the end for the length needed. You can also splice any piece of audio on the end and then turn down the volume for that clip. Other methods to make the system think there should be filler at the end can backfire when you make later changes and the sequence drops the fake filler. This happens if you splice a piece of video on the end and then lift it out. This is fast, but only a temporary fix. Any black or sound added to the end of a sequence ends up in the EDL if you don't remove it, which is why I prefer the method that does not affect the actual sequence.

If the sequence is longer than the first pass after the changes, then make sure the tape has been blacked long enough to accommodate the extra time. If not, change from insert mode to assemble mode in the Digital Cut window and on the record deck, put the timecode preset/regen switch in the regen position.

## IMPORTING AN EDL

After reading Chapter 7 you should be aware that an EDL is a terrible way to archive a sequence. However, there are times when you take a sequence to an online linear tape suite and assemble a project only to have it return back to the Avid suite for more changes! Now you need to get the EDL into your MC/AX and digitize it. This is taking a non-compatible format and translating it to something the Avid can understand.

Using EDL Manager 1.9 or later, you can open an EDL and either save it as an OMFI file or import it directly to the Media Composer if they are both running at the same time. Use the arrow icon on EDL Manager that is going back toward the Composer icon and the list should import into an open bin of your choice. This brings in a sequence and a series of master clips that are almost 24 hours long. The master clips are there for reference and are not really to be used. You can delete them because you will be getting new clips from decomposing the sequence.

Load the sequence into the record monitor and then use Remove matchframe edits from the Clip pull-down menu. Decompose the sequence and redigitize.

If your sequence does not import correctly, you must strip some of the complex information out of it. Open the EDL in Vantage (a word processor that ships with every system in your Utilities Folder) and choose Show invisible characters in Vantage. Make all unusual effects into dissolves, strip out comments and GPIs, and basically make the list as simple as possible while maintaining all the original character placement. Although EDL Manager 1.9 is more forgiving about missing spaces in the EDL, it is still best to leave everything in the original format. There are too many variables to guarantee that importing your list always comes in perfectly to the MC/AX but you should get the majority of it without a problem if you simplify as much as possible.

## VTR EMULATION

In the interests of completeness, there may be times when Media Composer does not control a deck but is controlled by an edit controller. This is useful if you want to tie Media Composer into an edit suite and use it like a disk recorder or just another deck. It could hold bits and pieces that you will be using over and over for a linear online assembly. You can also use VTR Emulation to output the sequence time to tape and slave other devices to Media Composer.

VTR Emulation requires a special cable that must be ordered from Avid and then you can set it to behave like any type of VTR that works best with your system. VTR Emulation defaults to a Sony PVW-2800. The H-Phase and the SC Phase also need to be adjusted in the output tool just like any deck, so that they match the timing of the switcher. If you adjust the H-phase and SC phase correctly, you don't get a giant horizontal shift or color change when you attempt to dissolve it with another source through a video switcher.

The methods in this chapter are relatively advanced and complicated, but they are created in response to the unexpected needs of the real world. Even if you can't remember all the steps involved in these different techniques, at least you know that these procedures are possible. There is a lot of flexibility when it comes to going from offline to online. This extra power means that you must spend a bit more time and go through more steps when you are working on a complicated sequence. If you find yourself able to always work at the highest resolution, you can save yourself much of this effort, but that may not be realistic. When the time comes that drives are so inexpensive that most people can work at high AVRs all the time, then all of these techniques will be as obsolete as knowing how to thread a quad VTR.

# 9

# Simple Troubleshooting

This chapter is devoted to keeping you off the phone and on the system. There are many everyday situations where just a little knowledge of troubleshooting can keep you going forward, give you a bit more confidence, and, maybe, keep your job. The first thing that can help you instantly is to read the Release Notes. This falls under the category of RTFM (Read The Manual—Please), but some people skip it because they just want to be up and running with the newest version no matter what. Stop, smell the roses, and read the known bugs. Avid is pretty good about listing what they consider to be the bugs you need to know about. Your definition of a bug and their definition may differ, but you will definitely benefit by seeing that, for instance, one small part of something you need to do all the time doesn't work under certain conditions. It also helps to know if that procedure has been replaced by something faster, better, and simpler. Even if you haven't memorized the Release Notes, they can generate a little thought bubble over your head if something seems familiar. So don't toss the Release Notes; keep them handy and even scan them quickly before you call Avid Customer Support.

Customer Support has improved so much in the last few years that average wait times as of this writing are under three minutes and Avid takes it very seriously, but for you that's not really the point. There is time involved in figuring out you have a problem, realizing you don't know how to fix it, telling the other people in the room maybe they had better get some coffee, and then dialing for help. Better to say something like, "Hmmm, did you know you are missing the active SCSI terminator?" and subtly implying, "Aren't you glad you hired me?" However friendly, competent, knowledgeable, and good-looking Avid Support may be, you want to avoid talking to them until you have a serious problem.

## ERROR MESSAGES

If you are not used to working on complex professional software, you may not be used to generating error messages. Write down the ones you see and, if they do not keep you from continuing, you can call Avid Support after the session and get the official explanation. Saying, "I got an error" isn't enough to let Avid Support help you figure out the problem. It may be something systemic or it may be operator error. But if it is a Fatal Error, you should be on the phone immediately if you don't know what caused it. After working on the system for a while, you will learn what causes the most common errors and if you keep a log of when they occur and to whom, you can identify the pattern that applies to your own system, facility, and way of working.

The real trick to error messages is in deciphering them. They may be colorful but essentially meaningless until you figure out what the function is that has gone wrong or what kind of pattern they follow. A section of the programming code always generates a specific error message when something goes wrong. Essentially, error messages tell you what happened, not why. This is because the same error could have been caused for ten different reasons. The computer cannot look outside of itself and say, "You have too many unrendered effects at AVR 77 and the wrong terminator on Media drive 4A and that last sound effect on audio track 6 just put me over the edge!" It will say "Audio Underrun" because what happened was that it could not continue to play every frame of audio and video through that segment of the sequence. Many times this is the only explanation that the computer can confidently produce for why it cannot play.

## THE IMPORTANCE OF CONNECTIONS

First, let's look at the basic world of connections. You may have enjoyed playing air guitar to Molly Hatchet, but air SCSI doesn't work as well unless it's connected to something and connected correctly. Connections are one of the first things to look at, especially if you have just moved the system. "I just moved my monitor and now it is broken," will lead most support reps to check your cables to the monitor.

Many of the connections to the computer have electricity running through them and connecting anything "hot" can cause a component to burn out. A good rule of thumb is not to change connections while the computer is running (not including audio or video to an external source like a deck).

Cables that are screwed in tight don't come loose so quickly. This seems obvious, but many facilities decide it is easier to have them not screwed in so anyone without a Phillips head screwdriver can move things and make changes quickly. Do yourself a favor and buy a screwdriver with one end Phillips head and the other end flat. Tighten everything on the back of the computer and anywhere else you can tighten things down. Of course, if you yank really hard, you still have problems because now you have loosened the computer board it was attached to or even ripped the wiring out of the connector!

Make sure that things are fastened down away from big feet and spilling coffee, but don't pack them away so tightly that you can't get at them to take a look. Get a few bags of little plastic tie wraps from your favorite electronics store and wrap cables together in logical groups. You have to cut the tie wrap to pull them apart at some point, but better to discourage the unauthorized and you can always wrap them again when you are done. The cost compared to downtime is negligible. Good engineers leave a little slack in all their carefully tie-wrapped suites so that any piece of equipment can slide forward enough to see around back. Large, easy-to-read labels that are also easy to understand make any phone call to Customer Support less embarrassing. "Is the video input connected?" "Uh, you mean cable VI649?" If you need to call your own video engineers at least, you will be sure that the problem is not software related. Also contemplate moving everything in the Avid suite a few feet away from the wall and mounting a small, clip-on lamp back there. This gives the system some more air if the room tends to get warm and will give you easy access.

## MONITOR CONNECTIONS

If you have an Avid Xpress, you already know the benefit of working with high-resolution monitors. More open bins (if they are small) keep you speeding along under certain projects. But more interesting is that when you are digitizing, you can have the waveform/vectorscope, the audio VU meters, several bins, and the console all open at once. On the other monitor of the Media Composer, the record monitor, you can get a huge amount of real estate for your timeline. Getting the Source/Record window with the two screens to get larger is a little tricky. You have to mess around with the cursor on the bottom right-hand corner until it changes shape and then drag the window so that it fills the screen horizontally. This positioning is saved with your user settings so you only have to do it once.

There are two common problems with monitors. The first is getting the RGB and sync cables connected wrong in the back. How can you tell? All your colors will be horribly wrong. The monitor may be completely black if the sync cable is not connected correctly.

The second problem occurs only on the 20-inch monitors and has to do with the termination of the monitor. Your signal will be washed out and way too bright. Many people foolishly live with this because they didn't expect the signal to look great in the first place! They accommodate it by cranking down the brightness controls on the front of the monitor. To see this, look at the white type of the splash screens as you launch the software. If the letters are a little too bright, check the termination. Look around back of the monitor, near the input connections, for the termination switch. It has an icon that says "75Ω" which stands for 75 ohms. Make sure your monitor is terminated since it is very easy to whack that little 75Ω switch while moving or unpacking.

## ADB CABLES

One of the silliest problems that seems to happen often is with the keyboard cables (ADB or Apple Desktop Bus). First, let me caution everyone who is thinking of pulling this harmless looking cable and reconnecting it while the system is running. You may have done this a hundred times but the next time you do it, you can fry your keyboard or even your motherboard. Any cable that has power running through it has the capability to create a power surge or damage sensitive electronic parts by "hot swapping" or changing the connection while the power is running through it. The ADB connects directly to the motherboard and the connection to the board is not replaceable. If that ADB connection fries, you need a new computer. If you don't know which cables have power and which do not, change all connections that are not standard video and audio only when the system is shut down. I know this is overkill, but simple and safe as a general policy.

The ADB cables definitely have power running through them. Unfortunately, ADB cables are always about an inch or two too short and so are under a certain amount of tension all the time. Many times just a slight pull is all it takes and your system appears to crash. You try all the keyboard reboot commands and nothing works and finally go to the CPU itself to hit the reset switch. When everything comes back on line, surprise! You still don't have control, because what really happened was the ADB cable had come just slightly loose.

One last consideration with the ADB chain is how many devices to attach in the chain. Many people like to use trackballs because they don't have a lot of space to manipulate a mouse. They also like the programmable buttons for repetitive or common functions. For those that don't like the feel of a trackball, there are programmable mice now too. And the most versatile device to connect to the ADB chain is the graphics tablet. Instead of suffering the effects of the mouse or trackball death grip at the end of the day, many people like the feel of a light, familiar feeling pen. Even if you find that you don't like editing with the pen, it is the only way to do graphics. Simple graphics, like fixing a scratch or a dropout, becomes faster and easier when you get away from the "painting with a bar of soap" feel of the mouse.

## SCSI CONNECTIONS

The ADB connection is basic and straightforward compared to the scariest of all computer connections — SCSI (Small Computer System Interface). The term SCSI Voodoo may not be completely foreign to you and for a good reason. Even though you may follow all of the complicated rules of SCSI or simplify your system to minimize them, you may still encounter situations that just don't make sense

## Basic SCSI Rules

There are some basic rules to keep in mind no matter what your SCSI configuration is. First, all media drives that are playing video at any resolution above AVR 3 should be connected to the SCSI accelerator card and not to the other SCSI connection that comes built into the Macintosh. That other connection, away from the other cards, should be used for devices like slower drives for graphics, backing up, removable media, scanners, or the external 3-D effects "pizza box" (the Aladdin). Avid is forced to complicate the SCSI situation because standard SCSI connections are not fast enough to pass the huge amounts of video and audio data at the speeds required for real-time playback at high resolution. Avid adds an extra card to the Macintosh called a SCSI accelerator. The SCSI accelerator is made by a company called ATTO and this is where you generally connect all your media drives. The more things you have connected at once, the more complicated your SCSI troubleshooting will be.

Second, keep the length of your SCSI connections as short as possible. The cables that ship with your drives are meant to be that short because anything longer than the maximum length of normal SCSI chain (Avid supports 18 feet) causes serious voodoo behavior. The length of the cable used for the entire SCSI chain must take into account the length of cable inside each drive. That can add up pretty fast at over a foot per drive. A common mistake is to take one look at the length of the cables that came with the drives and rush out to the electronics store and get the longest cable you can find. Avoid the temptation to use these and keep the cables short.

SCSI cables are very sensitive to twisting and must be handled more carefully than any other cable on the system. Just bending a SCSI cable back and forth a few times can significantly reduce its functionality. Don't strain, kick, stomp, or bite them.

There are lots of complicated rules about SCSI, especially when you are combining the newer wide drives with the older narrow drives. These three terms, fast, wide, and narrow, refer to the speed and capabilities of passing larger amounts of data through the SCSI chain. The drives themselves don't look all that different. Fast and wide drives (the is9 and is23 drives for example) are the only kind of drive supported for two-way striping to playback AVR 77. You can use the older four-gigabyte and nine-gigabyte drives for AVR 77, but you must use a dual ATTO card or two ATTO cards and stripe four of the same size drives together (two on each ATTO card). This is called *four-way striping* and is like having one really fast drive with four read/write heads. Four-way striping is a good way to continue using older drives, but if one drive fails, you will lose all the data on the other three drives too.

There are two types of SCSI on an MC/AX: regular, single-ended SCSI and SCSI-D or differential SCSI. The differential SCSI is for drive towers and storage expanders that allow you to connect many more drives than either the standard seven or eight and are capable of the much longer cabling lengths of 75 feet (compared to 18 feet for single ended SCSI). The important thing about a differential connection is that it has a non-compatible connector to keep you from accidentally connecting regular drives and burning them out because of the difference in the amount of power flowing through the cables. Don't confuse MediaDocks with drive towers. MediaDocks do not require the differential ATTO card or the differential cables.

## Termination

All SCSI chains must be terminated. Termination is the way the computer knows where the end of the chain is and which is the last device. It keeps the signal from bouncing back to confuse the computer with false signals. Technically, the chain must be terminated at the beginning and at the end, but on most Macintoshes (all the newer ones and all the ATTO cards), the termination at the beginning is internal. This means that you need to attach the terminator that comes with drives to the last drive in the chain. Always be sure to use an active terminator, the purple terminator that came with the narrow drives or the blue one that came with the termination kit for the wide drives. Don't use the generic gray terminators from the electronics store—they are generally not active.

## SCSI ID

The most basic SCSI rule is that the number of the SCSI ID, the number associated with this drive in the chain, must be unique to the chain. The SCSI ID is set on the back of standard drives with a pen or other pointed object; in MediaDocks it is set on the front panel. Before you attach any drive, make sure that its ID does not repeat a number already being used by another drive on that SCSI chain. The fact that this fundamentally important piece of information is a small number on the back of the drive again points out the importance of giving yourself access to the equipment after it is installed. How do you know which numbers are being used unless you can stick your head back there with a flashlight and read them upside down and backwards? The best way may actually be another piece of software—the Avid Drive Utility (ADU). This utility ships with every system and shows you the correct SCSI ID even if the ID number on the back of the drive is broken and displaying a wrong number! Of course, you need to be able to boot your system with the drives attached in order to use the ADU.

Getting the wrong SCSI ID may cause the system to not boot correctly. The Macintosh desktop may also indicate that you have many more drives connected than you really do. When there is a problem with the SCSI chain on startup, you may get the flashing question mark icon. This is because the Macintosh cannot find the startup drive with the system folder on the SCSI chain. If you are having a problem with SCSI IDs, however, do not try to change the SCSI ID while the system is running!

With narrow drives, you can only use the numbers 0 through 6 for a SCSI ID. A wide drive can use 0 through 6 and, if connected to a PCI Macintosh, 8 through 15 as well. Never, ever use SCSI ID 7 on any SCSI device since this is the number used by the ATTO card itself (which is also technically a SCSI device) or by the Macintosh. On the Macintosh SCSI port, avoid SCSI ID 0 because that is the ID used by some internal, system drives. Internal CD-ROMs are set at the factory to use ID 3. When you are adding and subtracting drives from any system, the ID is the most important factor in making sure the drives work happily together.

## Connecting Wide and Narrow Drives

We have already discussed the difference between narrow and wide drives and, when you upgrade to wide drives, you need to figure out how to connect them to the same SCSI chain with the narrow drives. The narrow drives must come last in the SCSI chain.

When connecting wide and narrow drives on the same chain, you need a special termination kit from Avid. Wide drives use a 68-pin connector and the narrow drives use a 50-pin connector. The difference is more than cosmetic, however, since there are more active pins inside the 68-pin connector and these extra pins must be terminated before connecting to the narrow drive. You get a new wide cable to go between the last wide drive in the chain and the first narrow drive, a small blue terminator/adapter that allows this cable to connect to the narrow drive, and a blue terminator to go onto the last drive in the chain. Use the standard wide cable (68 pin to 68 pin) to go from the Macintosh to the first wide drive. Then use the termination kit cable (also 68 to 68 pin) with a special blue terminator adapter (68 pin to 50 pin) added to it from the kit when you connect the wide and narrow drives together. Finally, use the special blue terminator on the last drive in the chain.

Cables to connect the wide ATTO card to the narrow drives (68 to 50 pin) are very different from cables that appear similar for connecting wide drives to the narrow drives (68 to 68 pin with a 50-pin adapter/terminator). Even though this narrow cable may connect, you will have nothing but problems connecting wide drives to narrow drives. It is a good idea to label these cables as for narrow use exclusively. Better to lock up the 68- to 50-pin narrow cables somewhere after upgrading to wide drives.

There are still many complications to the SCSI chain, but these are the basics and must be taken into consideration on an everyday basis. Even if you follow all these rules, you may find that a particular drive works on ID 4 but not ID 2. You may never find a good reason for this (although there is a reason). You may also find that the order of the drives in the chain makes a difference even though all of the termination is correct at the end of the chain. And above all, you may have to juggle extra devices on the Macintosh SCSI to find the best order and take into consideration that some devices come already terminated internally and must always go at the end of the chain, like the 3-D DVE. If you have two such devices that terminate internally, they are going to fight with each other until you eliminate one or figure out how to unterminate it. Sometimes this requires going into the guts of the device! Going inside an Avid drive nullifies your warranty so don't ever open the outside case on the drive (Avid can tell if you did).

## AUDIO CONNECTIONS

You also need to be concerned with audio connections. If you are doing anything with multiple audio sources, consider a small, inexpensive mixer. Check to see if your audio card is an Audio Media II or an Audio Media III card. As I pointed out in an earlier chapter, be sure to use the Avid-supplied audio cables going from any source, like a mixer, into the audio card. There is a built-in attenuator in the cables that makes sure that the level going in matches the level going out. You cannot compensate for the lack of these attenuating cables unless you have a mixer that outputs at -10 dB.

If you have self-powered speakers, connect the speakers to a power strip with everything else and turn them on through that strip. This is a good idea for most of the equipment since it lessens the possibilities that a piece of equipment will be left off accidentally. If the speakers are on and you still have no sound, check the Audio Tool (Command-1) to make sure you actually have audio playing back. Change the Text in your Timeline to show the media file name and see if it says Offline. If it gives you media file numbers in the Timeline audio tracks and you still can't hear any sound, then check your Pro Tools external hardware. Make sure that the sample rate is set to the same rate the audio was digitized at. This can be confirmed with a heading in the bin or by looking in the Console, which will clearly tell you that you have a sampling mismatch. If the sound was originally digitized at 44.1 kHz and the switch is set to 48 kHz, you will hear nothing. Also make sure you are set to .99 or 1.00 to match the original digitize setting as well. This switch can be hit by accident or when switching back and forth from a film project that may need the .99 pull-down rate. If you only need one type of sampling rate all the time, I suggest using a little piece of black gaffers tape to permanently hide these switches from accidental change.

## THE BLACKBURST GENERATOR

A blackburst generator (BB Gen) is like a synchronizing clock for video or audio. It provides a steady source of a perfect video signal: black. I highly recommend one of these even if it doesn't come with the system you ordered. You can connect it to decks, monitors, your video card, and the Pro Tools video slave driver to synchronize your audio. Many decks require a stable signal to their composite video input when they are playing back, especially the Sony PVW-1800. If you are using the component inputs and outputs, consider connecting blackburst to the composite video input permanently. It makes it easier to black tapes when you are not using the deck for anything else.

You should connect things as permanently as you can and design a system that requires as little connecting and disconnecting as possible. Consider patch bays, mixers, MediaDocks, networks, and removable media like Zip or Jaz drives. The less wear and tear you put on cables, the more reliable and long lasting they will be. A little more initial investment during the planning stages can positively reduce troubleshooting downtime in the future.

## STANDARD COMPUTER WOES

Even if you successfully eliminate the potential for problems with connections, there is still the potential for standard computer type problems. Unfortunately, these kinds of problems are not easily solvable by the typical MIS department, even if you are lucky enough to have one. This is one of the reasons it is a good idea to be able to take care of your own computer problems. Standard computer support personnel are going to be at a loss with the Avid problems unless they have been through some training for the specific requirements of high-resolution video.

### Extension Conflicts

An extension is a small piece of software that launches when you start the computer and runs in the background. They are necessary for the operation of certain software applications. The MC/AX has a handful of extensions that are installed automatically in the Extensions folder in the System folder when you use the installer disk. You can run Adobe Photoshop and After Effects and some other third-party programs that require their own extensions on your MC/AX computer without problems and, while many times this is efficient and practical, it is sometimes asking for trouble.

If you use programs that add extensions, when you use their installer, eventually you will have an extension conflict. Sometimes the conflict is obvious and you cannot boot the system and run the Avid software after you have installed something. That problem is pretty easy to fix. In fact, there is a program called Conflict Catcher by Casaday and Green, Inc. that systematically cleans out your extensions folder and puts extensions back in a way that allows you to quickly zero in on the culprit. You may find your life simplified by making several sets of extensions and labeling them for the particular functions they are used with. Have a set of extensions for graphics work, one for cruising the Internet and one that is stripped down for editing. The best set of extensions to create is one that allows you to work with the three or four most useful programs that you run all at once. Test to make sure there are no conflicts, strip the extension folder so that only those applications are supported and then save it as an extension set. (Back it up in case you are forced to reinstall your operating system.) Your system will run faster on less RAM and give you many, many fewer mystery crashes.

The most difficult extension problems to track are intermittent or related specifically to particular functions. They may occur only when you go into the digitize mode or only when you are making a digital cut. These may occur because some third-party application has changed some important function in your system. Another video application may reset the frame size of the video board or change the media drive firmware. There are many small, but important settings that may need to be reset after running another program on the MC/AX Macintosh. You may only be able to find these programs after they cause serious problems, but if those problems are intermittent, you may never connect the two as cause and effect. Simple is better and if you can afford another Macintosh to run third-party applications, you can reduce the amount of crashes when you are running the Avid software.

### Working from a Floppy

An organizational error that is unfortunately too common is trying to work directly from a floppy disk. Do not ever open a project, a bin, or a user setting directly from a floppy. The moment you remove the floppy, but continue to work with the bin, you will have very strange errors. The MC/AX periodically needs to save back to the original bin file on the floppy. Even worse, you will not be able to close the project or save the bin you are working on if it grows too big to store back on the floppy. All bins automatically save themselves when they close or when the project is closed. You must allow for the extra space if the bin has grown larger.

### Access to Original Software

One of the final solutions to difficult, intermittent, or unusual problems is to reload the software. Avid Support may ask you to reload the Avid software, the Macintosh operating system, or both. All users of the system should have access to these disks. A big mistake, commonly made, is to lock these disks up safely away from anyone who might need them, probably at 3 AM. Make sure the disks are the most recent and correct versions of both the Macintosh operating system and the Avid software. More problems are caused by loading an unapproved version of the Macintosh operating system than may be solved.

## Access to the Hardware

None of these techniques do you any good unless you actually have access to the Macintosh itself. Some installations have cleverly hidden the system away in another room or rack mounted it in a machine room. If you are completely forbidden to touch the hardware because of facility or union rules, then forget about it. You can just hand the phone to the appropriate authority. Otherwise you must, phone in hand, be able to look around back and see that all the cables are tightly connected, that the power to everything is on, and see what the disks' SCSI IDs are.

All of this means that you should look into getting a good engineer to set up your suite. Good engineers are worth much more than their salaries when they save you the embarrassment of your first several jobs going out the door with bad levels because you were monitoring the audio in the wrong place! Make sure, before the engineer leaves, that you have a thorough understanding of the cabling and get a wiring diagram you can refer to.

## Never Enough RAM

More RAM is always better, but there is one thing you must do after buying and installing (or getting a certified Apple Technician to install) your RAM. You must change the allocation of the application. Find the icon of the original Media Composer software and click on it once. Then choose Get Info (Command-I) from the File pull-down menu at the Macintosh Finder level. Change the minimum and preferred amount of RAM to take all of the new RAM you installed except for about seven to ten megabytes for small programs and enough for the Macintosh operating system to expand when it needs to. To be really sure you are not shortchanging the operating system, go to About this Macintosh under the Apple Menu in the upper left-hand corner of your screen while you are at the Finder level. See how much RAM the system is using now and add a few megabytes to it. Subtract that amount from the entire amount of RAM available and give it to the MC/AX. If you have a large amount RAM, well over 100 megabytes, and you don't have a huge amount of media online, over 200,000 objects, then you might want to leave a larger chunk available for a graphics program or EDL Manager.

## Starting Up from Another Drive

One of the first things you need to do if you have problems when you boot your system is to try to start up using another drive. It is easy enough to bypass all your extensions if you think that is causing the problem by holding down the Shift key during the initial moments of the startup. But if the problem is truly with the internal Macintosh drive, you can force the Macintosh to look for another drive with a system folder on it and see if you can boot at all. Once your system is up and running, you can try to track down the extension conflict or run a disk repair utility to try to fix the problem.

The Macintosh looks to see if there is a floppy disk with a startup folder already in the floppy drive before it looks to its own internal drive. This is where Norton's Emergency Disk comes in handy. It contains a stripped-down system folder and Norton's Disk Doctor, which you can then launch from the floppy and use to repair the internal drive. If you don't have an Emergency Disk, you can force the Macintosh to look for another drive before the internal drive, like the Disk Tools floppy that came with your system software (if it came on floppies). You can boot from a CD-ROM that holds the installation of the Macintosh operating system too. If you simply hold down "C" while booting, you will default to starting up from the CD-ROM. If you hold down Command-Option-Shift-Delete, you can boot from a CD-ROM, a Zip, or a media drive if they have a viable system folder. I carry a Zip drive and 100 megabytes of utility programs with me. I also carry a clean System folder that I copy from a properly functioning Avid system.

Again, having access to a clean Macintosh operating system and the CD-ROM of the original MC/AX application may be an easier and faster answer than really trying to pinpoint a problem. You can always go back and try to recreate the problem when you have more time and no one is threatening you with deadlines. In fact, an easy way to do a clean reinstallation of the Macintosh operating system is to hold down the keys Command-Shift-K (for klean, I guess) while rebooting and it automatically puts the old operating system in a folder called Previous System Folder. With version 7.6 and later, there is a checkbox for the clean install as Apple finally realized how important this operation is in a pinch.

## ROUTINE MAINTENANCE

By far the most important day-to-day steps you can take to keep your system running and happy are just routine maintenance. These techniques are so fast and easy that as a freelancer, I used to come in a little early on the first day of a project to run them on any new machine before I started.

### Rebuilding the Desktop

The first procedure is called rebuilding the desktop. Do not underestimate the power of this procedure. You are telling the Macintosh to look at every file on all the drives and build a new database of exactly where everything is. There are so many important pieces of software that the Mac must find instantly to run smoothly that having everything accounted for before you begin can prevent all kinds of little nasty surprises. When you start up in the morning, or if you are restarting because you have had a problem, hold down the Option and Command keys during the startup process. It takes longer to start up as the machine looks at everything on all the drives, but it is worth the wait in the long run. You don't need to do this every day, but if you have been moving, deleting, or adding large amounts of files, then it's like chicken soup: it can't hurt. You can rebuild the desktop using Norton's Disk Doctor, which we will discuss shortly. When in doubt, ask Avid Support about versions of any third-party programs that you use on your system for maintenance.

## Disk Diagnostic Programs

The next important step of routine maintenance, a disk diagnostic program, takes a little longer, but if run once or twice a week, maybe more often in a facility that never sleeps, it allows you to catch serious problems before they become fatal. Disk diagnostic and repair programs have been around for a long time and people may have had a bad experience with them and media files, but they are a lifesaver to run on internal drives. Norton Disk Doctor is the most popular and ships with the MC/AX. On the latest versions of the Avid Installation CD, it is in the Utilities folder with just the parts that you should use. In the past you may have received the entire shrink-wrapped Norton Disk Utilities package. If you did a full install of all the options, you installed a few extensions that can interfere with the operation of the MC/AX. All of this correct installation information, of course, was in the Release Notes, but . . . just install the Disk Doctor and leave the rest for a Macintosh that doesn't need to play back gigantic video files with real-time effects!

There have also been the rumors that you should never run Norton Disk Doctor on media drives. Part of the problem was that in trying to repair a gigantic media file, old versions of Disk Doctor occasionally left the file corrupted. More recently, the controversy was over the fact that Disk Doctor couldn't deal with striped drives. With Disk Doctor version 3.2 and later, that is no longer true. I run Disk Doctor on the media drives now but, with one exception. I never repair a media file. If Disk Doctor wants to repair header information and other minor directory problems, it can go right ahead, but I don't want it to mess with my 100-megabyte media files.

Speed Disk is a utility that Norton ships with Norton Disk Utilities which defragments, or reorganizes a drive so that there is no wasted space. It optimizes disk space by moving files around and, in a very orderly way, filling in gaps on the drive. This is great if you are concerned about space, but not so great if you are concerned about performance of gigantic audio and video files which must playback simultaneously. Avid has determined how to record files to the drive in such a way that video and audio stay together to maximize efficiency when the drive head is reading the file. That way, the drive doesn't have to work so hard at the high speeds that are necessary to keep it from underrunning. If you digitize audio and video to separate drives, this is not such an issue, but many people don't have that luxury. Note that if you put audio and video that are meant to run simultaneously on different partitions of the same drive, you are forcing the drive to work even harder. With the popularity of striped drives (and the repeated assurance that you will not lose half your drive space), digitizing audio and video to different drives is less of an issue. If you are really pushing the system, trying to get lots of real-time effects at high resolution for instance, and you are pretty organized, then by all means, split your audio and video to separate drives or even separate ATTO cards.

A new tool to help determine if the drive is really dying is DiskEx. This utility is available from Avid Support and ships with new systems. DiskEx exercises your disk by making it run through a series of stress tests to see if it will fail completely. If you run DiskEx and your disk does not fail then, you can confidently search elsewhere for the problem instead of waiting overnight for a new drive which will not solve your problem.

## TECHNIQUES FOR ISOLATING HARDWARE FROM SOFTWARE

Part of the trick of troubleshooting is to determine, before you go very far, whether the problem originates with the hardware or the software. If the problem is internal to the computer hardware, the video and audio cards, then there is very little a novice can do without many cautions and a bit of hands-on training. In early versions of the Media Composer, most systems were Quadra 950-based. The Quadra 950 was very easy to get in and out of and was relatively robust as a chunk of metal and plastic, but it had one drawback: Over time the boards would heat up and cool down and eventually work themselves just a little bit out of the slots. The video and compression boards especially were large and heavy and, if you had the Quadra sitting straight up and down, they would pull themselves out by their own weight. The best strategy if you are still using a Quadra, and hopefully you have upgraded it with a Power PC card, is to lie the computer flat on its side with the side door on the top.

If you really need to get into the Quadra, you should not be stacking drives on it either. If you suspect your board is loose, the only recourse you have is to reseat it. Gently pull the board out of the socket and push it firmly, but gently back in. But you must take serious precautions against electrostatic discharge (ESD) and potentially voiding a warranty, so this is not to be done lightly. I'll go into ESD later in this chapter.

If you suspect hardware-related problems, it is always worth a call to Avid Customer Support. If you have determined that the problem is not hardware related, then look to the software. This is the main reason that all Avid Certified Support Representatives are thoroughly trained to understand a wide range of uses of the system. There is nothing more frustrating than having a video engineer or some other technician called to examine the Avid Suite and have them shrug their shoulders and say, "Must be the software . . ." and leave.

## SOFTWARE PROBLEMS

There are two major areas where you could have problems in the MC/AX application: the project files, bins, and user settings, or the media itself. Try to separate problems with the media playback from problems with the media itself. Media can always be redigitized, but I'll bet you don't have a few extra video cards hanging around. The best way to tell if a particular problem rests with the media is to take a close look at it. Step through the problem areas frame-by-frame. That will show if the problem is there when the drives are not working so hard. Any corrupt images or crazy colors that are visible when you are looking at a still frame can be solved by redigitizing that shot. If you can't see the problem, you must go one step deeper.

Is the problem playback related? See if the sequence plays back without any of the media online. Go to the Finder level and dismount your drives by dragging them to the trash (the drive icons not the folders inside) or by changing the name of the folder on the media drives from 5.*x*, 6.*x*, or OMFIMediaFiles to anything else. I just add an x to the end of the name so I don't get into trouble trying to spell the correct name of the folder again when I am in a hurry. As discussed previously, any change to the name of this very important folder causes all the media inside to go offline. Now does the sequence play? If there are no errors, you know that the problem is related to the drive or the computer board. If you continue to get obscure errors (my favorite is BadMagic), then you need to look to the sequence itself.

There may be something corrupt in your bins, your project, or your user settings. This is so easy to test that many times it is one of the first things an Avid Support Representative asks you to do. Generally I suspect this particular problem when something that has been working fine all day stops working or features that should be available are suddenly not there. Create a new bin or project and drag the clips and the sequence you were working on into the new bins.

The next step is to remake your user settings. This should be a fast check to see if default user settings clear up the problem. If they do, then spend the time to recreate your settings. Better yet, call up an archived version of your user settings that has been hidden where no one can get to it. This may also be corrupted, but chances are pretty good that it is not. Again, make sure you have new user settings for every major change in the software. If you call up a really old user setting, it may be incompatible with new menus and functions in subtle, but important ways.

## AUDIO DISTORTION

A common audio problem occurs when you are working along and suddenly the audio starts to sound distorted. If you have heard this audio before, you know it was not digitized at levels that were too high. If you call up the Audio Tool you can make sure the digitized levels do not go into the red area of the VU meter. It helps to isolate this problem if you have digitized this material yourself and you know that the sound was also recorded correctly in the field. If you have access to the original tapes with the Media Composer, use a button called Find Frame. After pressing the Find Frame button, the system asks you to insert the original tape and cues up to the exact same frame you are viewing in the editing window.

If the audio levels are good on the tape and have been digitized correctly, but still sound distorted when you play them back in the software, you may have a corrupt file called the DigiSetup. This is a little file that is created automatically, is always there, and sometimes, when it is being updated, gets corrupted. The best thing to do is just track it down and delete it. DigiSetup is in the Macintosh System folder. You can find it with Find File or by opening the System folder and pressing the letter D. Later versions of the MC/AX make one for you automatically when you launch the application the next time. Otherwise, you need to launch a utility called Pro Tools Setup and just click OK. Another sign that DigiSetup is corrupted is if you can play audio normally for a few seconds, then it just stops or goes silent, yet if you stop and play that section again, there is audio. If the distorted audio is also slightly slowed down, like listening at 95% of speed, trashing DigiSetup is always the answer. If all this does not solve your distortion problems, you may need to look to the audio board itself.

## FONTS

Even though Apple has tried to make dealing with fonts as simple as possible, somehow it is still complicated. When you try to use the wrong kind of font in the Title Tool, it may look absolutely terrible although all the other fonts look fine. The main complication has to do with what kind of font you have and where is it located. Standard TrueType or Adobe PostScript fonts installed correctly should be available to use in the Title Tool. You should have a copy of Adobe Type Manager (ATM) loaded to deal with the Adobe PostScript fonts.

The confusion comes when people download a font from the Internet and install it or take a single-sized font and try to resize it. These are two different problems. The first situation, a font of unknown origin, generally means that it is a bitmapped font. It is meant to be only one size and if you try to change it in the Title Tool it quickly looks chunky with ragged edges. You should really discard this font; it is of no use to you with the MC/AX.

The second situation is similar, but results from a different cause. When you are using Adobe PostScript fonts on a Macintosh there are two parts to every font, a screen font and a printer font. You may find that one of these, usually the screen font, has not been installed correctly on the system. This means that you no longer have a font that rescales for screen display. Time to go back to the original installation disk and reinstall the font. Sometimes good fonts just go bad. They become corrupt and they must also be reinstalled again. You should always have access to the original software, especially fonts, in case you have to reinstall your Macintosh system software from the CD-ROM.

Incorrect installation may account for fonts not showing up at all. All fonts must go into the Fonts folder in the System folder. You can simplify this by dropping them onto the closed folder icon for the System folder and the Macintosh asks you if you want them to go into the Fonts folder. You say yes and it does the rest after you restart. The problems occur when you just drag them into the open System folder and drop them inside that open window. Then they are not put away correctly and are loose in the System Folder.

Another problem with fonts is that too many of them take up unnecessary amounts of RAM. If you want to slim down your required RAM for the operating system so that you can give the RAM to another application, then removing fonts helps as soon as you restart. Another way to simplify font selection, but complicate your life, is to use font suitcases. There are many programs that allow you keep a particular set of fonts together and load them all at once so you have just the ones you need.

## MEDIA MANAGEMENT

Another day-to-day concern mentioned before is the management of media on the drives. You may think that you are efficiently deleting all media as you finish each project, but you may be surprised to find all sorts of odd bits of precomputes and imported graphics and so forth floating around on your media drives. The best way to keep a handle on this is to delete media from the Media Tool and not through individual bins. That way you see the project from a big picture point-of-view and can evaluate on a project-by-project basis what must go and what must stay. It also keeps you from accidentally deleting media from another project just because you dragged a duplicate of someone else's master clip into the bin you are now deleting.

The real problem, not only with having many unnecessary small objects on your drives slowing down performance, is that the drives may accidentally become too full. You should always keep a minimum of 50 MB on any partition or any drive completely free. There are all sorts of files, like the Mediafile Database, which must occasionally get larger to accommodate the changing nature of the media on the drives. Don't worry about defragmenting media drives, but overfilled drives corrupt media and may crash and take everything with them.

## SERIAL PORTS

Printers and modems can be connected to your system, but they take up valuable serial port connections. This may be the best reason to use Ethernet for printers since it frees up a serial port so that you can use it for deck control. The modem port and the printer port are identical and interchangeable. It is just that certain software expects to find the modem connected to the modem port and the printer to the printer port. If they can't find the right device, tell the software to look at another port either in the Chooser or in the software itself.

Sometimes the serial ports get stuck in an open or closed setting and appear not to recognize that there is a device connected to them. This might happen if you had a printer or a phone connection problem and the last thing you did with that device did not shut it down or turn it off properly. You may notice this in the MC/AX when you have lost deck control, although everything else is set up correctly. You have the Serial Tool under Tools and you can change which port you are using for deck control or change the port to "None" and change it back to the one you want.

If the Serial Tool doesn't clear the serial port, you must either use a utility called CommCloser that ships with the Avid software or zap the PRAM (pronounced pee-RAM). The PRAM is parameter RAM and is a part of the computer memory that holds important information that keeps things like the clock going when the system is shut down. PRAM remembers important settings so that when you boot up everything is the way you left it. Sometimes you must purge this memory to clear the stuck serial port. TechTool, a piece of freeware from Micromat Computer Systems, is the best utility for this because it zaps the PRAM quickly. Otherwise, you must restart and hold down the P-R-Command-Option keys and wait for the restart chime to sound eight times. This can be pretty tedious if you have a lot of RAM because the restart takes so much longer. You then just restart and reset all the control panels that may have changed back to their default state.

The most important settings to put back are memory settings. Modern Memory Manager must be on. With a Quadra, you must turn the 32-bit addressing back on in the Memory Panel. All newer systems should check the Memory control panel and turn virtual memory off. If you have a Quadra with a Daystar Power PC upgrade card, you must turn it back on through the Control Panel and shut down completely (not just restart). It can take some time to both go through the procedure and to reset all the control panels, so zapping the PRAM is generally considered a last resort.

## VERSION NUMBERS

You should always have, either in the back of your head or written down someplace accessible, the version number of almost everything associated with your system. First and foremost is the version number of the MC/AX software. You should know it down to the last digit because every small change in the software has a reason. Avid puts out what they call the "gold" version of the software, the most tested and most stable version they can achieve within the time allowed before it must be released. Version 6.5 is an example of a gold release. Release dates, as with all software companies, are based on complicated interrelationships all lining up at the same time. If, after the gold version is shipped, some features didn't make it into the software although they were planned, there may be other releases with an extra decimal point. Release 6.5.2 is an example of that. There may be some procedures that were not tested and appear to have problems, so there may be another revision called a patch release. A patch release is meant to fix one or two small problems. This would be version 6.5.1v2. The important thing to remember about all these releases is that you might not need any of them except the gold version. The other versions have not been tested as thoroughly as the gold release because of the importance of getting out a fix in a timely manner. Most of the time this is not a problem but asking for a patch release if you don't absolutely need it, pushes the envelope unnecessarily.

It is always best to have all the systems in your facility running the same version of the MC/AX. Most versions are forward compatible, but not backward. This means that a bin created in version 5.5 opens fine in version 6.5, but not the other way around. Once a bin has been opened in the higher version, it has been converted forever to that version. By just opening the bin once in version 6.5, you may not be able to open it in version 5.5 again. You have to export the bin full of clips as a shot log and import it into the earlier version. Unfortunately, this does not work well for sequences. There is a bin converter program that goes between versions 6.5 and 5.5.1, but it is best to not depend on such complications. If you are forced to change systems in the middle of a job, always insist on the same version of the software or higher.

You should know the version of the Macintosh operating system you are using. Some versions are not approved for some models and you should make sure that your Macintosh can run the latest software before you install it. For instance, Apple did not approve the Macintosh Operating System 7.6 for the 9600. Oops. You must use an earlier or later version.

Always make sure you know the latest version Macintosh operating system Avid has tested. If at all humanly possible, try to get all your Macs to run the same Macintosh operating software. Keeping everything interchangeable is a valuable goal and should not be complicated by an MIS person or especially enthusiastic editor who wants to put the newest on their machine as soon as it is available.

Version numbers also carry over to the hardware. Each of your boards has a revision number and that should be considered during troubleshooting. You may have an old version of a board or a version with a known conflict. You can check the revision of your hardware in several places. If you have MC/AX running, look under the Hardware Tool to see each individual board. Also there is a utility called Avid System Utility that comes on PCI systems and Slot Looker for Nubus that allows you to get more detailed information about each board. This is the better answer if you can't actually launch the MC/AX. If Avid System Utility or Slot Looker can't see the card at all, calling it an unknown card or the slot is empty, then that tells you the problem is with the card or the way it is mounted in the slot.

Every utility that ships with your MC/AX has a specific version number. The ones that ship are the ones that are meant either for that hardware or that software. Again, grabbing a version from another system just because it is newer may get you into trouble. Certain versions of the Avid Drive Utility, for instance, are designed primarily for four-way drive striping. The rule of thumb is: If the new version has a tangible improvement, completely compatible with your system and supported in the Release Notes, only then should you load it onto your Macintosh.

## ELECTROSTATIC DISCHARGE (ESD)

The sneakiest and hardest problem of all to diagnose is one that is very easy to create: damage related to electrostatic discharge (ESD). You may not realize it, but the human body can store and discharge frightening amounts of static electricity. Shuffling across the carpet with a relative humidity of 10% generates 35,000 volts! Compare this to the smaller, faster devices that are necessary to achieve the kind of performance necessary to keep the system running happily and they have a range of susceptibility of several hundred volts.

You can zap a component with static electricity by touching the outside of an ungrounded device or, more probably, by opening the device to do some simple troubleshooting. You may be asked to remove or add RAM or to reseat a troublesome board. Anytime you are going to open a case, be concerned about voiding a warranty or causing ESD damage to the sensitive components inside.

The key to touching anything inside the system is to be grounded. Being grounded ensures that any build-up of static electricity is channeled off to a ground and dissipated. This is best done by wearing a grounding wrist strap and connecting it to a ground or to a metal component inside the computer. Most important, make sure the computer itself is grounded. The best way to ensure this is to plug the computer into a grounded outlet. An even better solution is to plug the computer into a grounded power strip and then shut the power strip off. That way you are not supplying power to the Mac while you are working on it.

The scariest thing about ESD is that it doesn't always kill, sometimes it just maims. A board or a RAM chip that receives a substantial shock may not fail right away. It may not fail for days, weeks, or months. It may start to show intermittent behavioral problems that cannot be isolated. These are absolutely the worst kinds of problems to troubleshoot because they may not occur for long periods and may not be of a type that points to any one component. It may be the Macintosh itself that received the shock and no matter how many boards you replace, it does not solve the problem. This is why ESD should be treated so seriously and any time you handle a component or open the Macintosh, you should be very, very careful.

## CALLING CUSTOMER SUPPORT

If, after all these precautions and general maintenance, you must still call Avid Support, at least take heart in a very short, best in class, wait time. To make it go even faster, have certain answers prepared since almost all support calls start with the same basic questions. Know your versions, Mac OS, Macintosh, and Avid model. Be able to describe what you were trying to do when the problem occurred. This is especially crucial for video engineers who have not taken the time to learn the software. The editor describes the problem and the engineer cannot explain it to the support rep in enough detail. What is the exact wording of the error? Some errors are pretty obscure like "missing a quiesce." Write it down and don't fake it with "something was missing, I think." When did the problem begin? Right after you put in the new RAM? Pull the RAM out. Does the problem happen every time you perform a particular operation or is it really random? Can you repeat it? The simplest answer to any error that seems random is to shut down the Mac and restart.

The other very important thing you need to give to Customer Support is your system ID. If you don't have a system ID you may not get any support! That is the way Avid determines whether you or your company has a valid support contract. Not something you want to discover at 3 AM. The system ID can often be found by just doing a Finder search for "sys" or launching the utility Dongle Dumper, but if you can't launch the computer that won't help much. Write it down or use Dongle Dumper and print it out.

We can't deal with all the techniques, potential problems, and error messages in this short space. If you want to know more about your system there are courses in troubleshooting and a whole curriculum to become an Avid Certified Support Representative. You don't need to be a technician to keep your system running happily most of the time, but you should have some basic knowledge of what is going on "under the hood." Good maintenance routines and a healthy dose of caution are two necessary items when dealing with sophisticated and complicated systems. Keep it simple and you will be rewarded with fewer steps and less stress when you need to troubleshoot.

# 10
# *Introduction to Effects*

The effect capabilities of Media Composer and Avid Xpress are surprisingly deep. Even with the more limited layering features of the Avid Xpress, wonderfully complex and professional results can be achieved very quickly. At this stage in the technology you can have two streams of video, one real-time effect, and an uncompressed title all play at once in real time (along with 8 channels of audio). The Media Composer offers from 8 to 24 tracks of video and the Avid Xpress offers 4. Even with fewer video tracks you can nest, collapse, and mix down to extend and expand any system's potential.

With the cheaper hourly rates to create effects that are normally unattainable in the low- or mid-range budgets, some rendering seems quite an acceptable expense of time. For those productions that are used to paying $700 per hour, these effects are not going to truly compete. The last time I checked, time was still money. But what is the difference in price and the difference in value with the higher quality of a high-end compositing station? With versions 7.0/2.0, the lines between cheaper/slower and expensive/faster will further blur and it will soon be hard to ignore Avid effects as a viable possibility for all but the high-end needs.

There is much to be covered to deal completely with effects; too much for this book. But there are some basics that can get you past the beginner stage. There is nothing like the experience of a hands-on class and Avid offers two of them, one for an introduction and one for more advanced features and techniques. There is also a fine CD-ROM created with two levels of instruction for beginner and advanced levels. This chapter can only hint at some of the techniques you will discover with enough time to experiment.

Anyone who is serious about effects should own the 3-D option. This not only dramatically expands the repertoire of tricks, it gives you more real-time effects. As you become more confident with effects, you will also become serious about nesting. Nesting is the feature that gives you more levels of video layering than you could ever practically use except for the densest of graphics sequences. This chapter will discuss nesting later after the basics. Nesting gives you incredible power with an extra level of complexity.

## TYPES OF EFFECTS

There are three kinds of effects: real-time, non-real-time and conditional real-time. Real-time effects can be handled by the compression and video boards without rendering as long as there are only two sources of video playing at the same time or there is only one effect at a time. If more streams of video or multiple, simultaneous effects are desired, then rendering is necessary. The exception to this is the new ability to play real-time and animate uncompressed titles over one layer of real-time effects. This just recently became possible with existing hardware with the release of versions 7.0/2.0.

Real-time effects have an orange dot over the effect icon when unrendered. The good news about real-time effects is that they can take advantage of a "fast render" scheme that allows them to render at between eight and six times real time. The faster CPUs cannot benefit from this faster rendering time because much of the rendering happens on Avid's special rendering boards. But a faster bus, the part of the computer that all information must squeeze through going to and from the boards, can make a significant difference. The PCI bus system (9500, 9600, and up) improves rendering times over a NuBus system (950, 8100, and AMP).

A non-real-time effect is too complex to be dealt with so quickly and sports a blue dot once it is in the sequence. It must be rendered by the Macintosh itself and may take 20 times real time or more to render. Clearly, if a real-time effect does the trick, it is preferable and design should take this into consideration. An imported graphic with an alpha channel is a real-time matte key, but a matte key made by using the matte key effect in the Effect Palette is a non-real-time effect. This is because the imported matte key has a third stream, the alpha channel, that does not change over time. It can be loaded into RAM and cheated into being "a run length encoded" source, not a real video stream. A matte key created with the matte key effect from the Effect Palette assumes that the alpha channel will animate over time and so needs the full playback and computing power of a real video stream.

But there are certain effects that are too complex and have too many moving, complex edges that must be anti-aliased for the fast render hardware to cope. Occasionally, these blue dot effects have an extra choice for rendering: draft quality or highest quality. They also have an extra icon that toggles to show either a stair step (aliased draft quality) or a smooth diagonal (anti-aliased highest quality). The difference between draft and highest quality for rendering time can easily double an already longer rendering time for this effect. So if you are working at low resolution for offline purposes or you are still changing the effect sequence on a regular basis, you will not want to spend the time on highest quality. You can change the quality when you render the individual effect, or when you "render in to out," draft, highest, or as set in each effect.

A conditional real-time effect is an effect which, under certain conditions, may switch from being real time to an effect that must be rendered. It is not a non-real-time effect, however, and the colored dot is green not blue. The green dot indicates that the conditional real-time effect can still take advantage of the fast rendering scheme. They are not nearly so slow to render as a true non-real-time effect. Without the 3-D option, the user must choose between real-time keys or real-time wipes. Whichever effect is not chosen is a conditional real-time effect. Real-time effects also become conditional when there are two effects on the same master clip. If you put a dissolve on a color effect, the dissolve becomes a conditional real-time effect. Defaulting to the transition effect for the conditional effect ensures that usually it is the shorter effect that needs to be rendered.

The one other time an effect becomes conditional is when there are two 3-D effects that have completely different shapes; say one is a page turn and the other is a water drop, within five seconds or less of each other. One effect will need to be rendered.

The conditional status does not occur if two real-time effects are stacked on top of each other on multiple video tracks. They do not turn green, although something must be rendered. The system does not look vertically for effects, just horizontally, on the same video track.

When layering effects on top of each other vertically, the material on the top track always has priority. This is not a true multi-channel digital effect device in the way most people think of standard DVEs. It is more like having many single channel devices. Each video track can be considered a separate channel of effects and, with nesting, much more than that. But it means that the separate video tracks do not interact with each other because they are each like separate sequences. This modularity allows quick exchanges of shots when it comes time to modify an effect, but it means that if moving objects are going to change their layering priority on the screen, then the lower object must be moved to a higher video track.

## EFFECT DESIGN

Technically, AVR 75 does not support real-time effects on striped narrow drives and, if you are not using versions 7.0/2.0, then AVR 77 does not support real-time effects at all, even on striped fast and wides. But people do it all the time. Every now and then they have a title over a complex background at AVR 75 or 77 and they get an underrun error. This happens when the drives and the computer cannot keep up a constant stream of video without dropping a frame and system stops playback and gives you this error. The error may say audio underrun or video underrun, but it is most likely a problem with the complexity of the video that is causing the audio to underrun (although simplifying the audio may fix the problem).

Good effect design tries to achieve the most spectacular effects with the simplest use of layers. The fewer layers, the fewer problems with trimming and rendering. Simpler design means most of the time you can modify faster because it is easier to figure out what is affecting what. If you can do something in fewer tracks, it looks better and renders faster.

## RENDERING

Rendering effects can be significantly reduced by using some basic strategies. In general, it is only the top track that must be rendered. Whenever you render an effect, you are rendering a composite of everything below. Unless you want to play the tracks below by themselves, by moving the video monitor down to lower tracks or stripping off the very top tracks for multiple versions, you can leave these lower tracks unrendered.

What confuses people is that many times there is not just one track available to render as the top track. The beginning of the show may have a complicated layering section that has ten layers. Then most of the show may not go above track three and the end has five tracks. How can you set it up to just render a top track and walk away? The secret is using a fake effect, the submaster, and putting it on the video track above the effect sequence, just to have something to render.

To make sure that the submaster goes on the absolute top track, when you create a new video track, hold down the Option key. Then you are given a choice about which track you actually want to create. If you create video track 15 and place your submaster there, it is likely that you will not really need all of the tracks below. If you continue to add video tracks, then they are added incrementally, track three, then track four, etc., and leave your top track unaffected. Highlight the top track that contains the submasters, mark inpoints and outpoints around the entire sequence and render in to out.

A more complex approach is to render with a submaster and then plan to leave a top track in real time if you can. The titles always look slightly better in

**Figure 10.1** Applying a Submaster

**Figure 10.2**  Rendering the Top Track

real time and you can fade them in and out with keyframes instead of dissolves. There is a new Fade Effect button in versions 7.0/2.0 that adds keyframes for fade-in and fade-out to imported matte keys as well as titles. Keeping a Picture-in-Picture (PIP) or 3-D Warp in real time on the top track means that you can continue to make changes right up to the last minute while the lower layers stay rendered. Those last minute changes to the top layer effect may be the change of the shot for something else, for a second version of the effect or just a final tweak of the move itself.

## SOFTENING

As layers are added to a sequence and there is more detail in the image, some extra compression of the image is required to assure playback at that AVR. This is the drawback of working with lossy compression and is the main reason many people spend the big bucks for uncompressed effects. Minor amounts of compression are handled invisibly, but when the level of complexity gets too great, you need to soften or simplify the effect.

You may get a message that an effect must be softened while it is rendering. If at all possible, never soften an effect on purpose. Not only is it unacceptable for finished quality work, but it also drops out of the fast render mode to do the softening. I always set the Render setting to "Do not soften." This means that at a later date I might get an underrun error message because the drives can not keep up with playing back the level of complexity of that image. If I get the underrun error, I do almost anything else before softening. You can reduce the amount of audio tracks with an audio mixdown or render all the real-time effects around a trouble spot to help simplify that part of the sequence.

The softening message you might get during rendering is a little conservative. Avid engineers are always planning for the worst-case scenario of a maximum amount of audio tracks, rubberbanding audio, and highly complex real-time effects when estimating the ability to play back a complex image. You may never get near that level of demand on your system in normal work. This

**Figure 10.3** Render Settings

means that if your sequence is simple, but the individual effect is complex, you have some wiggle room to push the machine beyond the prescribed limit for a short time. Some people like to know if their effect is too complex. They may have problems playing back later, so they leave the Render setting button at "Ask Me." Then they set the render limit number above 100% to maybe 120%. This means they are only alerted if the effect is seriously over the supported limit of complexity. I always put this setting to "Don't soften" and have almost never had a playback problem because of effects. However, if you have non-striped, slower drives, this may occasionally cause a problem with high AVRs.

## KEYFRAMES

Almost all effects can be manipulated by keyframes. The only exceptions are the color effect and a few other segment effects like flip and flop. Keyframes are the method to change an effect over time and so there are always at least two keyframes: beginning and end. At the first keyframe, certain values about position, shape, or color are entered and the settings change to match the values on the next keyframe. The change, if there is any, is smoothly interpolated between the keyframes. Keyframes can be added in the effect mode on the fly while playing and hitting the keyframe key (the apostrophe key) or, on older versions, by mapping the locator button to a function key (this doesn't work in the wireframe mode with 3-D). Keyframes can be copied and pasted, dragged by holding down the Option key, and, in version 6.5 and later, moved with the trim keys. If you want the effect to just hang on the screen, with no motion, then you can copy and paste the same settings between two keyframes.

## Keyframes and Speed

The speed of any effect depends on the time between the keyframes and the amount of distance the effect must travel on the screen. If the keyframes are close together in time and the distance on the screen is great, then the effect appears to be fast. The overall length of the clip in the sequence determines the distance between the first and last keyframes. If the effect is put onto a very long clip, it takes the length of the clip for the effect to end.

Many times there is a situation where you want the effect to move rapidly and then stop with a longer duration clip. This requires three keyframes. The first two will be close together in time to provide the proper speed of the move. Then there is no change between the second and third keyframes as the shot plays out. The only exception to this is when using the Spline parameter with the 3-D Effects option. Spline always tries to smooth the path between keyframes—even if they are not moving. With Media Composer, you can turn Tension up to 100 in the Spline parameters so that there appears to be no movement between two keyframes that do not otherwise change position.

If you trim a clip and make it longer after adding an effect, the effect gets longer too. If you trim a clip with an effect shorter, you may create an impossible situation. If there is a keyframe on every frame and then you shorten the effect, where do the keyframes go? No one knows. You need to adjust this effect by hand after shortening because the results will be unpredictable.

## SAVING EFFECT TEMPLATES

After all this work making the effects just right, you can save them as effects templates so they can be easily applied over and over again. Any effect can be stored in a bin by clicking on the effect icon in the upper left-hand corner of the Effect Mode palette, and dragging and dropping the effect icon in the bin. If a bin that holds an effect template is open, that effect will be available in the effect palette. The effect template can also be dragged back from the bin and applied to the Timeline. When you apply the effect, it looks slightly different from the original effect unless the new clip you apply it to is exactly the same length as the original clip.

If the Option key is held down when dragging the effect icon to a bin, then the effect template is saved with the video. This is an "effect with source" and can be edited into the sequence like a master clip. If you add an effect to a title, then just saving the effect template for the title is always "with source." If you want the title effect template to just be the keyframes alone, so the effect can be applied to another title, then hold down the Option key when saving it. The trick is to remember that, when saving a title template and a standard clip template, the Option key has the opposite effect.

## ADD EDITS

If an effect cannot be manipulated with keyframes, it can be split into sections using the Add Edit button. By splitting one effect into multiple effects, each one can be manipulated separately and then recombined by a dissolve which, unfortunately, must be rendered. This is most useful for a color effect that can be used to change a color over time with add edits and dissolves. Due to hardware restrictions on older systems, the color effect may occasionally have a flashframe at the end if the effect is a radical change from the normal master clip. This flashframe can be fixed by rendering the entire color effect.

Adding add edits adds extra keyframes to an effect sequence. If you have an effect with two keyframes and you split it with an add edit, then you have two effects with two keyframes each. This can affect the motion of an object if you have turned on acceleration. Acceleration is the smoothing of motion with a slow-fast-slow speed change over time that approximates the physics of natural motion. Adding an extra keyframe causes the effect to slow to a stop at the new point. If there is motion you should only apply add edits to split an effect at a point where a keyframe already exists.

## NESTING

Nesting is the most complex, but the most powerful characteristic of effects with the MC/AX. So far we have discussed building effects vertically which, depending on your model system, may be limited. Once you understand nesting you can expand the amount of tracks dramatically. Nesting involves stepping into an effect and adding video tracks inside. More effects can be added inside the nest and then you can step into those. It is a way of layering multiple effects on a single clip, but also much more. The only real limitations are how long you want to render and how much RAM you have.

There are two methods to view a nest with Media Composer and one with the Avid Xpress. With Media Composer you can apply an effect and then use the two arrows at the bottom left of the Timeline to step in or out. Once inside the nest you can no longer hear audio and cannot tell how many layers you have stepped in. But you can focus on that level alone and work on it like it is a separate sequence. Within that layer you can add as many new video tracks as your model allows.

The other method to view nesting is used on both Media Composer and Avid Xpress. Using the segment arrow to double-click on a segment with an

**Figure 10.4** Traditional Nesting

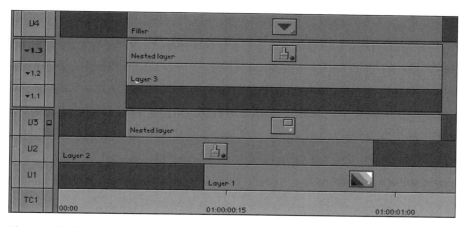

**Figure 10.5** The New Nesting Display

effect, the tracks in the Timeline expand to see all the layers inside at the next level. Continuing to double-click layers inside the first effect reveals those tracks as well. Tracks can be patched and edited in this mode and the audio can still be monitored. It is a little easier to understand all the effects going on because the display is more graphic.

This mode of viewing nesting can frustrate people who open it up by accident and then are confused about what they are looking at. You can always close the expanded view by double-clicking on the original track again with the segment arrow or, in Media Composer, Option-clicking on the down nesting arrow. The view can be turned off in Media Composer with a checkbox in the Timeline Settings dialog box.

**Figure 10.6** Timeline Settings

## Autonesting

Nesting this way implies a certain order of assembly: Apply the outer effect and then step inside. You must apply the PIP and then step in for the color effect. But real life doesn't always work this way. Many times the nest is a secondary thought, used well after the first effect is in place and rendered. In this case there is autonesting. Select the clip with the segment arrow and then Option-double-click an effect in the effect palette. The second effect does not replace the first effect, but it covers it. This builds the layers from the outside instead of stepping into the effect and building them from the inside.

All of these methods are for adding multiple effects to a single clip, but they are just as useful for adding one effect to multiple clips. If there is to be a color effect that covers an entire montage, then it is a waste of time and energy to put a separate effect on every clip. What happens when the effect must be changed? Now you need to change one, turn it into a template, and apply it to all the others. But there is a faster way that uses autonesting.

If you Shift-select multiple clips in the sequence with the segment arrow and then Option-double-click on the effect in the effect palette, the effect autonests as one effect that covers all the clips. Adjust the one effect and all the clips are changed. If you want to change one of the clips and replace it with another shot, just step inside the effect and make the edit. You can also step inside the multiple-clip effect and add dissolves or other transition effects. You must render these inside effects, but you can leave the outside effect in real time to allow for future

**Figure 10.7** Autonesting

**Figure 10.8** Displaying a Nest

changes. Of course, if you render the outside effect, it will create a composite of everything inside.

## Collapsing Effects

You can nest an entire effect sequence into one effect after it has been built in a more traditional fashion using vertical layers. If keeping track of all the video layers becomes tedious, you can collapse them into a single layer. In order to nest effects, you must have an outside effect at the outermost level. Collapsing places a submaster effect over all the layers and nests them inside. Selecting the area to be collapsed by marking in and out and highlighting the desired tracks, you can collapse all the tracks except the top track which, like a title for instance, you can leave in real time.

There is no way to really uncollapse an effect segment. One must step inside, mark an inpoint and outpoint around the entire segment, turn on all the video tracks (except for V1 if it is empty), and then use the Clipboard button. If you use Option-Clipboard while inside the Collapse, then the layered segment goes directly to the Source window. Cut the segment back over the top of the collapsed effect in the Timeline.

Rendering a nested effect is simple; just render the top, outside effect, the submaster. This leaves all the effects inside unrendered, but it is sufficient to play as long as you are monitoring that outside effect.

You don't need to leave the top track of a collapse as a submaster. You can replace the submaster with another segment effect by just dragging and dropping from the effect palette. You can replace the submaster with a mask or color effect. Again, just like the strategy of leaving the top track in real time in a traditional, multi-track effect, if you step into a nest and render the top track inside the nest, the outside effect can continue to be manipulated. You can keep tweaking the effect on the outside of the nest if the dissolves inside are rendered.

## Collapse vs. Video Mixdown

Although the collapse feature is excellent for simplifying complex effects sequences down to one video track, a collapse can still potentially become unrendered. If you

**Figure 10.9** Leaving the Top Track in Real Time

are sure that an effect sequence will never need to be changed, match framed back to an original source, or used for an EDL, then you can use video mixdown. Video mixdown (under the Special menu) takes any section of video between marked points, whether it has effects on it or not, and turns it into one new media file. This new media file has no timecode (which timecode would it use if you had fifteen layers?) and breaks all links to the original media. This is why match frame and EDLs are no good.

Video mixdown should only be used for finishing and for something that you will be using as a single unit over and over, like the graphics bed for an opening sequence you use every week. Once all the effects are renderd, then a video mixdown is as fast as copying the media and the rendered precomputes to another place on the drive, an insignificant amount of time. If the effects are not rendered before the mixdown, then they will be rendered first as part of the video mixdown process, but don't forget to count on the rendering time in your calculations. So the workflow encourages rendering first and video mixdown later when everything is signed off on.

A video mixdown will significantly improve the performance of the Avid during long sequences with lots of effects. Instead of forcing the computer to "build the pipes" for lots of complex effects with many short media files, it just needs to find one master clip. This means snappier reaction time when you press play. Always make a copy of the sequence before you overwrite a mixdown over your timecoded original sources so when the client changes their mind you will have a sequence to fall back on. Video mixdowns are very powerful and time saving for a wide range of purposes, but don't use them for offline if you plan to redigitize or make an EDL!

## 3-D EFFECTS

All 3-D effects come from one effect, the 3-D Warp, or you can promote titles and imported matte keys. The titles and imported matte keys can then be animated around the screen in real time. Having the extra piece of hardware gives you both real-time wipes and real-time keys.

The 3-D Effects option and "m" AVRs are completely incompatible. This is because to enable four screens to play at once for hardware multicam, the "m" resolutions are at a much reduced frame size. The 3-D Effects option must be disabled before working with this unusual frame size. This is accomplished by launching the application and holding down the "F" and "X" keys. You will be given the option to enable or disable the 3-D.

There are clearly more things you can do with the 3-D effects other than just spin and tilt images. You also have Stamp and Clear which uses a real-time buffer to stack unlimited layers without any rendering or loss of quality. These stamped layers must be added while the sequence plays in real time;

they cannot be rendered and must fade out all at once, but the possibility for titles is very powerful. Remember to clear a stamp buffer or the effect will pop back onto the screen after you have faded it out. And the ever-popular trail effect can be added to any object that moves by taking advantage of the same buffer used for Stamp and Clear. Trails also are real time only; render the trail and it disappears.

There is also corner pinning which allows you to fit four corners of an object so that it matches the edges of another object. It is not quite morphing, but it allows you to put images inside TV monitors, picture frames, or the like. The animator's tool for realistic motion, Spline, is also useful for getting fast, professional results for complex motion paths.

There are so many things you can do with the 3-D Option that I helped write a one-day course just for that. This is truly an area that calls for personal experimentation. Just taking some of the shapes and using them to warp images into interesting graphic backgrounds requires parameters you must discover for yourself.

## PAINT AND ANIMATE

Version 7.0 includes significant additions to the capabilities of intraframe effects. If you consider editing to be basically an interframe process, working between frames, then painting on a frame is intraframe. There is a full palette of familiar choices for anyone who has used third-party painting programs. Brushes can be changed and areas of the image can be blurred, color corrected, and generally affected like any standard paint program. Multiple layers of paint effects on the same frame are possible, but unlike the other, single-image paint programs, you can easily change the effects over time with keyframes. You also save significant time and trouble by keeping the entire process inside the single application. No exporting, painting on separate frames, and reimporting anymore. Although some programs, like Fractal Design's Painter allow overlapping of multiple frames with an onion-skin-type interface to reproduce what cell animators use, you cannot say, "Start on this frame and expand the effect until this frame." Interpolation between keyframes is the beauty of working in a video-based program for paint.

You can also isolate parts of the frame, say the sky, and draw a matte shape around it to make it a deeper blue. The ability to draw on an object and create control points to adjust curves with bezier handles and move the edge over time means that almost any part of an image can be manipulated separately from the whole. By creating points that change over time, any smooth, even motion in a shot can easily be followed with just a few keyframes. If the motion is jerky or unpredictable, you need more keyframes to adjust the control points. If the motion is truly difficult to follow or you are doing dozens of motion following

**Figure 10.10** Effect Editor: Interpolation between Keyframes

**Figure 10.11** Creating Control Points on an Object

effects, you may want to use another program that allows you to automatically track specific pixels over time, like Avid's Illusion.

The Color Effect has been improved to allow quick adjustment of contrast range and gamma using eyedroppers. Pick the darkest point of an image with the eyedropper and it changes to the blackest possible values. Choose the brightest part of the frame and that becomes the brightest possible values so that an image that was dull because of a narrow contrast range can make use of the full range of levels available. You can also quickly take illegal levels and bring them to within broadcast specifications for luminance and black level using the adjustable clip levels.

## AVX

A whole new range of choices is now available with the creation of the AVX (Avid Video eXchange) plug-ins standard. This is an interchange format so that other third-party effects companies can easily modify their existing product line for use as an Avid plug-in. These video-based effects are much better suited to use in a video-editing program than effects created specifically for a paint program.

We have just begun to see some of the fantastic effects coming from third-party AVX plug-ins creators. One already very popular is the Ultimatte plug-in. The amount of control over keying with colored backgrounds and the excellent edges with basic default settings means you can finish special effects projects shot over blue or green screen on the Avid.

## UNCOMPRESSED TITLES

Higher quality titles are available in versions 7.0/2.0 with the uncompressed choice for titles. Edges and blockiness are completely eliminated when titles are uncompressed. There is also the ability to run an uncompressed title on V2 and have unrendered, real-time effects on V1. This is because you are not running three streams of video, just two streams and an uncompressed PICT file. With the ability to copy and paste from a word processing program, instantly apply a custom style, and do rolls and crawls, it is becoming faster, simpler, and more feasible to use the Title Tool for large amounts of text. However, for truly complex manipulation of type, you still want to work with a program that can use vector-based graphics like EPS files.

The ability to layer, paint, and use plug-ins has given the Avid editor a whole range of tools and looks that were never available before in one package. It also makes certain effects, which were out of range for small budget programs, easily accessible. The trend in effects these days is slowly moving away from the "wow!" and to more subtle, invisible applications. But there is still a great need for the ability to turn any piece of video into a graphic or give it a graphic look. Editors and designers will always continue to experiment and push the technology to the limits and, with the tools now becoming available, nonlinear editing with the Avid will continue to generate new and fascinating high-quality effects.

# 11

# Nonlinear Video Assistants

With the adoption of new technology has come the blending of post-production responsibilities. Producers and writers become offline editors, offline editors become online editors, and the difference between film and video begins to blur. One thing that this shift has created is many new people who can edit well, but who don't have the inclination, ability, or time to get involved in the technical requirements.

This opens up the possibilities for an entirely new position: the nonlinear assistant. There have always been film assistants and their responsibilities are pretty clear. They handle all the day-to-day requirements of film handling, organization, and preparation for the editor. There have also been video assistants in the past although their roles have changed through the years and have, at times, been eliminated all together. It is inconceivable to lack a film assistant on a major feature, but many high-end production companies operate quite well without video assistants.

Assistants are very important if the design of the post-production facility is focused around a central machine room. The editor initiates the communication via an intercom system and tapes are changed, set up, and dubbed by this voice on the other end. Occasionally, that assistant is in the same room and can speed up the editing process by doubling as a sound engineer or a character generator operator.

The elimination or devaluing of the video assistant makes it harder for young people to break into the business. Since the film assistant on a nonlinear project may work a second shift while the editor works throughout the day, there is less opportunity for the interaction between master and apprentice. With tense clients who are paying large sums per hour for time-critical work, the production company that puts an unknown or untested quantity in the driver's seat is taking a risk. They may lose their client forever to the competition or may have to discount the session to appease them.

How then does one break into this business? There are no second video assistants like there would be in film, so what is the entry-level position? Many times it is whatever the facility needs at the time; a dubber, a receptionist, even a courier. However, with the advent of nonlinear editing, the new choice becomes the nonlinear assistant. In large and busy nonlinear post-production facilities there may be one nonlinear assistant per shift and three shifts. The entry level then becomes the graveyard shift and eventually the day or evening shift, where they can discretely observe skills that keep the editor in demand. The job doesn't really require the ambition to become an editor, but the people that gain the most from the nonlinear assistant position are those that need to know these subtle skills to move on to the next level.

What responsibilities should such an assistant be expected to perform? Much of the knowledge they need has been covered in this book. In fact, many editors perceive many of the techniques in this book as something only an assistant would perform. Others see it as required knowledge before starting a job! It is when a facility desires such a specialization of labor, either for personnel or billable reasons, that the assistants have the most value. They must perform all of the functions and have all the knowledge required to keep the systems running. The post supervisor instructs the assistant on all the requirements to keep as many jobs running smoothly as possible. The administrator or supervisor sees the big picture and the assistant performs the tasks.

These important daily tasks include digitizing, media management, basic maintenance, backup, and output. Anything that is required to prepare the suite for the editor and the smooth transition from one project to another is appropriate for the assistant. Each facility has their own set of requirements, but mastery of all these skills can make someone very valuable to any busy post-production facility.

## DIGITIZING

Digitizing also implies following logs, creating bins, setting levels, and understanding drives. The logs are handed off with the understanding that the marked takes or, possibly, everything should be digitized and the master clips named according to the description in the logs. Bad logs means bad bins unless the assistant knows something about the specific job and is given the freedom to create better master clip names. Each take must be assessed as to whether video or audio only should be digitized, thus maximizing space. Digitizing way too much by digitizing video for voiceovers or room tone will fill the drives prematurely and cause careful estimates to be off. On the other hand, editors may prefer that, if the drive space exists, everything, even incomplete takes, be kept. Don't assume that just because there isn't enough drive space that certain shots must left out. It is the assistant's responsibility to find the drives, connect them, and digitize everything if required.

The bins can be named based on tape name and the editor determines where to put the shots based on content later. Levels, as discussed previously, are crucial to the finishing stage. Distorted audio and blown-out video can come back to haunt the project. All assistants should know how to set levels and monitor them. They may not be perfect levels, but they should be as close as possible to require only a few last-minute tweaks.

## DRIVES

Understanding drives is crucial to making sure the digitized video can be played back. There is a setting in the MC/AX called Drive Filtering under the General setting. If this is on, only drives capable of playing back the selected AVR are available. Unfortunately, when people use non-Avid drives they must disable this setting all the time since, when the MC/AX does not recognize the firmware loaded onto the media drives, it assumes the drive is incapable of higher AVR playback. This is not always true, but it is safer than assuming every drive can play back every AVR. Not paying attention at this stage can mean having to copy huge amounts of material to the proper drives. It is possible to digitize to a Zip drive by accident with Drive Filtering turned off! Generally, however, if there is a mismatch between AVR and drive capabilities, there will be an error about either the video or audio overrunning its buffers.

If you can, split the audio and video to separate drives, *not* separate partitions of the same drive. The drives should be named so there is no confusion between what is a new drive and what is just another partition on the same drive. Never fill the drives too full. Leave a minimum of 50 MB free on any partition. Versions 7.0/2.0 includes the ability to split a single media file across multiple drives after it hits a certain length, like over thirty minutes. This will simplify digitizing long clips. They will no longer need to be broken apart by hand to fit on two drives and end up confusing everyone. The problem will lie in separating the two drives with the same clip. Like striped drives, splitting the clip over multiple drives means both drives must be present all the time or the clips are unreadable.

The assistant needs a thorough understanding of SCSI principles. There are many times when fast and wide drives must be combined with a narrow drive. Be sure you can recognize which is which and have the right terminators and cables. Chapter 9 discussed these problems in detail. Unless your facility has installed drives on a very fast network like fibre channel or SSA, you will always be moving drives and checking SCSI IDs. You need to be comfortable with Media Docks or RMAGS. All drives should have unique names so they can be moved to any system and still be identified.

Understanding drives also means understanding how to resuscitate an ailing one and when to call it quits and get another. The number of drives returned to Avid with nothing wrong is astounding. If you can get a drive back to full health in an hour or two, how does that impact the production schedule compared to waiting for the morning rush delivery? What are the replacement policies of your non-Avid drives? Learn how to use all the drive utilities. What is the latest version of the Drive Updater, the program that reloads the drives' firmware, and what are its benefits or drawbacks? Perhaps for your situation the latest Drive Updater would be a disaster. Don't always load the newest thing until you know all the ramifications for your individual needs. AvidDrive Utility should be your most important tool here. Use it for mounting, repartitioning, and, as a last resort, a destructive read/write test.

There are important differences between maintenance on a media drive and the internal Macintosh drive. Using the wrong version of Norton's Disk Doctor (only 3.2 and later are approved for striped drives) or SpeedDisk on a media drive can ruin your weekend.

## MEDIA MANAGEMENT

All the media on all the drives are under your jurisdiction. Knowing what to keep and what to discard is both incredibly important and commonplace. Media Mover is the most important tool here. It gives each project its own media folder on the media drives. Locking and unlocking media files and deleting precomputes are all regular procedures to be familiar with. Know how long it takes to copy media from one drive to another. Delete through Media Tool or drag entire projects to the trash in their media folders. One of the most common calls to Avid support concerns the deletion of precomputes; your knowledge will keep you off the phone and your company humming at the maximum drive capacity.

Understand the network and how it helps you to move media efficiently. Become the network expert if you can and research the possibilities. Figure out how to improve throughput and how to reduce the number of drives that must be moved.

## BASIC MAINTENANCE

If you can, take a Macintosh support class or the Avid Basic Troubleshooting class. If you have some talent at the technical side, consider becoming an Avid Certified Support Representative (ACSR). Always have a floppy that you can boot from and run disk-recovery programs. Be ready to strip all unneeded extensions from a system and know which ones are the absolute minimum to run your specific system. Be ready to do a clean reinstallation of the system software. Between projects do everything you can to make all the systems as similar as possible.

If you aspire to become an editor, there are people who will hold your technical expertise against you. They think you cannot be technically proficient and a true artist. You may have to work a little harder to prove them wrong. It only takes a couple of success stories where you save the system *and* the project before employers see the value in an editor who reduces their own downtime.

## BACKING UP

Consolidating is the most important feature for backing up. Some people have enough time and tape to back up everything in a project, but more likely, you will be forced to decide what to keep and what to discard. Learn all the variations for consolidating the final sequence and focus your efforts on only the media needed to recreate the sequence.

What happens to a project when it is finished? Create a database that can both retrieve the project and the individual bins. If you print out the bin for each tape and include that with the tape, you have a paper archive when all else fails.

## OUTPUT

All forms of output are important and critical to the next step in the project. The digital cut may be the master or the approval copy. If it is the master, the levels must be perfect and if it is just a VHS, it must contain all the video tracks of graphics and all the audio, mixed or direct. Do you have a Media Reader that can burn in the sequence timecode to the digital cut?

The EDL must be rock solid and at least spot checked for accuracy. A cut list should be scrutinized at every cut because the stakes are so much higher and the chance for adjustment at the next stage more remote. Make as many versions and as many disks as you have time for. Expect that, when the online assembly finally comes, someone may ask for something more. This means that if the request is for video-only EDLs, make a few with audio too just in case. Make several printouts to cover yourself, save time, and help the online editor if there is a mistake at their session. Everybody makes mistakes and, if you have an original copy of the EDL on paper, you can isolate the mistake more quickly as something done wrong after your hand-off.

Talk to professional sound studios. What are the most common mistakes you are likely to make when handing off media and projects to an Audiovision or Pro Tools session? What is right for one house may be wrong for another—maybe because the other facility handles your files wrong! Document all the steps you take to prepare the files and label everything clearly. It really is part of the job to prevent other people from making mistakes.

## REDIGITIZING

Does every shot need to be redigitized? It is possible to make an EDL with media offline. If the changes requested only pertain to the open and close of a sequence, you might be able to get away with only digitizing those sections and leaving the rest offline.

If the need for redigitizing is because the levels were set wrong, then the saved settings for that tape must be deleted before redigitizing. If some of the footage needs to be digitized in black and white, make sure to set the Compression Tool to monochrome. Even more important, change it back when done! The third monitor shows you a color image all the time because it is monitoring what the signal looks like before it is processed. How many times have people walked into an Avid suite during digitizing, looked at the third monitor, and said, "Hey, that doesn't look so bad!" So it does not reflect whether you have left monochrome on or are digitizing at the wrong AVR.

## BLACKING TAPE

It is desirable to know everything about connecting and operating video decks. However, this information is beyond the scope of this book, but it would include knowledge about reference signals, signal termination and loop through, front panel editing, input choices, and blacking tape. After spending an afternoon with a well-known, much-decorated, documentary editor I asked about his deck connections. "What's a BNC?" he said. Although he had a great attitude toward the new technology, there was quite a learning curve involved in making him self-sufficient.

If you are not supported by staff engineers, I highly recommend that you learn how to clean the video heads. Oxide flakes off videotape and sticks in the tiny gap that video heads use to read the video signal. If you have a regular staff of technicians, then *don't touch*, but make sure the video heads are serviced on a regular basis. If no one is regularly cleaning the heads on the video deck, you should take on the responsibility as routine. Cleaning the heads with a proper head-cleaning kit once a week during heavy usage is not a bad idea and if you are using a consumer deck, of course, there are head-cleaning cassettes. Consumer quality tape loses oxide faster than professional tape (*The English Patient* was cut using S-VHS and a Film Composer). There is a significant difference between digital and analog cleaning procedures so make sure you know the deck you are working on. Digital decks do some self-cleaning. Do *not* clean with alcohol. Alcohol leaves a gummy film when it dries. Use a freon substitute and a wipe that is recommended by the manufacturer. Many people like TekWipes. Some rental facilities cover the edges of the VTR top cover with a seal so they know if it was opened and tampered with. Check with the rental company before breaking the seal.

For connecting Betacam SP decks, or any decks that have both composite and component input, I recommend looping the blackburst signal out of the reference input to the composite input. This serves several purposes, but the most important is that you will never accidentally use the composite input for your master tape digital cut. It is always a black signal. The other benefit is that you can change the position of the switch on the front panel of the deck that changes the deck input from component to composite and then begin to black tape without disconnecting or reconnecting any cables.

If you buy tapes by the case, then a good practice is to black them all whenever there is any downtime. By turning down the audio inputs and switching over to composite video input, which has been connected to a source of black, you can black tapes at a moment's notice throughout the day. Be sure to turn the audio inputs back up (or pop them back into the preset position) and throw the front panel switch back to YRB (component) before you start a digital cut!

Use the four switches under the front panel of the Betacam deck and set them to DF (drop-frame) or NDF (non-drop-frame) if in NTSC, "internal," "record run," and "preset." This ensures that the timecode is generated from the internal timecode generator and is not waiting for an external signal. The timecode only increments (runs) when the tape is recording and the starting point is wherever you preset it to start, no matter what other timecode is on the tape.

The ability to preset a starting number for timecode is achieved by manipulating the buttons on the outside of the front panel. Each deck is slightly different, but the Sony decks require first pressing the Hold button and using either the "+" and "-" buttons or the search knob to both choose hours, minutes, seconds, and frames and what those numbers will be. Finally, the Set button is pressed and, although the number in the LED does not show it yet, when you begin recording, it starts incrementing from the desired preset number.

Other switches to check on a Betacam SP deck are on top of the front panel (as opposed to on its face or hidden by it), where the record inhibit switch is also located. The 2/4 field switch can sometimes be the cause of field-based problems when digitizing graphics from tape. You should always digitize everything with this switch in the four-field position (NTSC). Also, if you are trying to lay four discreet channels of audio to the deck, you'll need to switch the front panel audio switch that says 1/2 and 3/4. If it's on 1/2, then the audio sent to 1 and 2 is automatically laid to 3 and 4 if those input levels are up. Many people use channels 3 and 4 on the Betacam SP because of the superior signal-to-noise ratio. They may want four discreet channels from the Avid to the deck so the stereo music can go to channels 3 and 4 while the narration and sound bites go to channels 1 and 2. A final note of caution: Tracks 3 and 4 are recorded helically, alongside the video, and if you make a video insert edit on a Betacam SP master you will wipe tracks 3 and 4 for the length of the edit!

There may be more specific requirements for assistants at your facility, but these should cover the basic skills to make you useful from day 1. All markets are slightly different with different terminology and different kinds of clientele. Every chance you get to watch the editors work should be snatched up, not just to see how they use the interface, but what they are doing with it. Observe the way creative ideas are thrown around, accepted, rejected, modified, and experimented with. This is always more important than the equipment being used.

You have an advantage over assistants in the past because, if you can get permission, you can take the opportunity to cut your own versions of the scenes, commercials, or segments when the suite would otherwise be empty. You can't mess up the film or add wear to the tapes just dissecting what the editor did and trying a few variations of your own. If you are lucky you can get the editor to check them out and offer suggestions. A recommendation by the editor may move your career along faster than anything else you can do.

Positions change and so does technology. If you stay flexible, always willing to learn something new, you will always stay valuable to your employer. This book outlined some of the more important techniques to keep in mind with the Media Composer and the Avid Xpress, but by no means all. To keep up with the speed of change takes more effort and more research than you might have expected, but stay focused on the most important aspect of the technology, the story telling, and you will never be out of date.

# Index

ABVB (Avid Broadcast Video Board), 86, 128
  serial digital, 133
  video levels in, 133–34
ACSR (Avid Certified Support Representative), 204
ADB (Apple Desktop Bus)
  cables, 48, 166
  Extender, 48
Add edits, 192
Adobe After Effects, 104
ADP ports, PowerBook, 11
After Effects, Adobe, 104
AIFC, 21
AIFF (audio interchange file format), 129
ALE (Avid Log Exchange), 7, 14
Alpha channels, 91–92, 98–99, 153–55
Animate, and paint, 197–99
Aspect Ratio, 89–90
Assistants, nonlinear video, 201–8
  backing up, 205
  blacking tape, 206–8
  digitizing, 202–3
  drives, 203–4
  maintenance, 204–5
  media management, 204
  output, 205
  redigitizing, 206
ATTO cards, 48
Audio; *See also* Sound
  connections, 170
  distortion, 178
  inputting four channels of, 142
  keeping, 24–25
  levels, 141–43
  monitoring, 31–32
  for QuickTime, 102–3
  scrub, 31
Audio Media II (NuBus) cards, 141–43
Audio Media III (PCI) cards, 141–43
Audio Tool, 141

Autocaching defined, 121–22
Autonesting, 194–95
Avid
  administration of, 47–76
    auto-deletion of precomputes, 56–57
    AvidNet, 72–73
    backing up media, 67–69
    backing up projects, 70
    between sessions, 57–58
    changing project names, 66–67
    compression, complexity, and storage estimates, 54–56
    Consolidate, 58–61
    copying on finder levels, 61–62
    custom columns, 63
    deciding what to delete, 62
    electrical power, 49–50
    empty space, 57
    ergonomics, 50
    Fibrechannel-Arbitrated Loop, 72
    Lock Items in BIN commands, 64–66
    Media Database, 52–54
    media management, 50–52
    media relatives, 63–64
    MediaShare, 71–72
    miscellaneous options, 69
    networks, 70–71
    relinking, 74–76
    room design, 47–48
    servers, 73
    sneaker net, 73–74
    user settings, 70
    using common sense, 76
    using creation date, 62
  codec, 97–98
  continued development, 3
  EDLs, 107
  Technical Support, 44
*Avid Hardware Installation Guide, The*, 49
Avid Video Tool, correcting adjustments in, 138–39

Avid Xpress
  Mac, 3
  Media Composer and advantages over, 127–28
AvidDroid, 29
AvidNet, 72–73
AVRs, 66, 78, 105
  mixing AVR 12 and AVR 77, 149–50
  picking right, 128–30
    digitizing, 129
    editing sequences, 129–30
    films using multicam resolutions, 128
    NuVista+ video card, 128
    setting audio levels, 129
AVX (Avid Video eXchange), 199

Backing up, 43–44, 205
  media, 67–69
  not to another drive, 44
  projects, 70
  software, 44
Bars
  color, 157
  field, 135–36
  SMPTE, 134
Batch digitizing, 16–17
BB Gen (blackburst generator), 170–71
Betacam
  components, 132–33
  SP, 11, 108
Bins, organizing, 15–16
Blacking tape, 206–8
Blackouts, 50
Burn-in-timecode, 13

Cables, SCSI, 47
Cables, Apple Desktop Bus (ADB), 166
Caps lock keys and scrubbing, 31
Cards
  ATTO, 48
  Audio Media II (NuBus), 141–43
  Audio Media III (PCI), 141–43
CCIR (International Radio Consultative Committee), 92–94
CD-ROMs (compact disk, read only memories), 101–2
Changes
  planning for, 25–26
  positive, 1
Channels, alpha, 91–92, 98–99, 153–55
Cinepak, 102
Client is always right, 5
Clip Menu, 9
CMX controllers, 114

Color bars, 157
  importing, 157–59
  signals, 138
  and tones on master tapes, 157
Colors
  corrections
    scenarios, 22–23
    and sound mixing, 20–23
  correctors, 131
  measuring devices, 130
  safe, 96
  space conversion, 85–88
Columns, custom, 63
Command keys, 34–35
Command-I (Get Info), 55
Commands, Lock Items in BIN, 64–66
Common sense, 76
Compact Pro software, 44
Complexity, 54–56
Composite signals, 132, 136
Compression, 54–56
  spatial, 102
  temporal, 102
Computers
  make decisions easier, 5
  monitors, 131
  woes, 171–74
Connections
  audio, 170
  importance of, 164–65
  monitor, 165
  Small Computer System Interface (SCSI), 166–70
Console, 79
Consolidate tool, 58–61
Control panel, remote, 139
Controllers, CMX, 114
CPUs, multiple, 56
Creation date, 62
Customer support, calling, 183–84
Cut, final, 159–60


DAs (distribution amplifiers), 137
DAT (digital audio tape), 25
Databases, creating, 15
Date, creation, 62
DAW (digital audio workstations), 129
DD (double density) disks, 113–14
DDRs (digital disk recorders), 122
De-interlacing, 95–96
DeBabelizer, 96, 99–101
Decks, Betacam SP, 108
Decompose, 127–28, 144–45
Delete, deciding what to, 62
Design, room, 47–48

Desktops, rebuilding, 174–75
DF (drop-frame), 12–13, 207
DigiBeta decks, 11
DigiBeta masters, 11
Digital Cuts, 55
Digital project never really finished, 17
Digital videotape, 108
Digitize Mode, 55
Digitize Tool, 54
Digitized images, 56
Digitized tones, playback, 142
Digitizing, 202–3
  batch, 16–17
  fixing mistakes, 61
Digitizing scenarios
  decomposing without timecodes, 150–51
  drive space and offline media
    automatically offline, 147–48
    handles and new master clips, 147
    putting sequences back together, 148
    splitting sequences, 145–46
    using media mover to take video offline, 146
  low disk space variation, 149
  mixing AVR 12 and AVR 77, 149–50
  recreating titles, 151–52
  replacing graphics, 152–53
Disks
  DD (double density), 113–14
  diagnostic programs, 175–76
  HD (high density), 113–14
Distortion, audio, 178
DIY (do it yourself), 21
DLTs (digital linear tapes)
  archiving material, 68
  and backup software, 44
  cost of archiving, 25
  dealing with, 67
  not a foolproof way of backing up media, 69
  restoring and archiving projects on, 57
DOS (disk operating system) format and Sony, 114
Dragging, speeding up, 35
Drives, 203–4
  connecting wide and narrow, 169–70
  starting up from another, 173–74
  Zip or Jaz, 44
Drop-frame vs. finished masters, 13
Dubbing
  and tape wear, 10–11
  with timecode, 108–10
Dupe reels, 121–24
DVEs (digital video effects)
  devices, 137
  generators, 116

EditCam media, 6, 133
Editing
  flow, 17–19
  Media Tool for, 42–43
Edits
  add, 192
  assembling, 155–56
  option-add, 41
EDLs (edit decision lists), 11, 14, 24–25, 105–26
  Avid, 107
  defined, 115–16
  good, 105
  importing, 160
  Managers, 110–12
  simplifying, 118
  software, 110
  tape names in, 110
  third-party programs, 106
Effects
  add edits, 192
  Avid Video eXchange (AVX), 199
  design, 187–88
  introduction, 185–200
  keyframes, 190–91
  nesting, 192–96
    autonesting, 194–95
    collapsing effects, 195–96
  paint and animate, 197–99
  rendering effects, 188–89
  softening, 189–90
  templates, 191
  3-D, 196–97
  translating, 116–17
  types of, 186–87
    conditional real-time effects, 187
    non-real-time effects, 186
    real-time effects, 186
  uncompressed titles, 199–200
Electrical power, 49–50
Empty space, 57
Encapsulating defined, 97
*English Patient, The*, 206
Ergonomics, 50
Error messages, 163–64
ESD (electrostatic discharge), 176, 182–83
Ethernet, 71
Exporting, and importing, 85–104

Fast forward and rewind, 33
Fetch Button, 36
Fibrechannel-Arbitrated Loop, 72
Field bars, 135–36
FileMaker Pro program, 15
Files, PICT, 118, 157
Find Media Relatives, 63–64

Finder levels, 61–62
Finishing, 24
Fixed Q tables, 55–56
Floppies, formatting, 112–14
Folders, OMFIMediaFiles, 52, 74
Fonts, 178–79
Footroom defined, 133
Force to Fit, 89–90
Formats, 90–91
  long, 77–83
  PICT, 90–91
  prosumer, 11
  RT-11, 112–14
Formatting floppies and RT-11, 112–14
Fractal Design's Painter, 197
Frames, importance of last, 36
Freerun defined, 7–8

Gefen Systems ADB Extenders, 48
GIGO (garbage in, garbage out), 107
Graphics
  relinking, 153–55
  replacing, 152–53
  soundbites and concern about, 19
  tablets, 91
  television, 91–92
  and video, 117
  working with, 20
  workstations, 91–92
GVG (Grass Valley Group), 110, 112

Handles, 143
  and new master clips, 147
Hardware
  accessing to, 173
  isolating from software, 176–77
HD (high density) disks, 113–14
Headroom defined, 133
Heads, cleaning video, 206
Hue adjustments with composite signals, 136

Ikegami EditCam, 6
IMA (Interactive Multimedia Association), 102
Images
  compressed, 55
  digitized, 56
Importing and exporting, 85–104
  Adobe After Effects, 104
  color space conversion, 85–88
  de-interlacing, 95–96
  importing requirements, 88–92
    alpha channels, 91–92
    Force to Fit, 89–90
    format, 90–91

  Maintain Aspect Ratio, 89–90
  resolution, 90
motion video, 96–103
  audio for QuickTime, 102–3
  Cinepak, 102
  Media Cleaner Pro and CD-ROMs, 101–2
  PICT sequences, 99–101
  QuickTime, 97–99
Open Media Framework Interchange
    (OMFI), 103–4
  RGB vs. CCIR for import, 92–94
  safe colors, 96
Inputs, choosing, 131–33
Interlacing; See De-interlacing

J-K-L keys, 29–30
Jaz Drives, 44, 69
Jog, mouse, 32–33
Jumping precisely, 34–35

Keyboards
  custom, 28–29
  using, 27–31
Keyframes, 190–91
Keys, Command, 34–35

Levels, audio, 141–43
Linear online
  be prepared, 115
  dubbing with time code, 108–10
  dupe reels, 121–24
  EDL defined, 115–16
  EDL Manager, 110–12
  formatting floppies and RT-11, 112–14
  getting ready, 112
  matchback, 126
  multiple layers of graphics and video, 117
  offline cut in online, 108
  PAL lists, 125
  preparing for, 105–26
  preparing sound, 107–8
  printing lists, 124
  real-life variations, 114
  settings, 119–20
  simplifying EDL, 118
  Sony and DOS, 114
  sort modes, 120–21
  sound levels, 118–19
  source table, 124–25
  tape names in EDLs, 110
  translating effects, 116–17
Linking, 74-76; See also Relinking; Unlinking
Lists, PAL, 125
Lock Items in BIN commands, 64–66

Locks, sync, 38–39
Logging
  by hand, 11–12
  defined, 6–7
LTC (longitudinal timecode), 139

Machine, talking to, 1–4
Macintosh, assumptions understanding, 4
Maintain Aspect Ratio, 89–90
Maintenance, 204–5
Management, media, 50–52, 179–80, 204
Master clips, 74–76, 147
  consolidating, 59–60
Master tapes, 157
  color bars and tones on, 157
  timecodes on, 154–55
Matchback defined, 126
Material, non-timecoded, 45
MC/AX (Media Composer, Avid Xpress)
  bins, 7
  compatible with EditCam media, 6
  creates new master clips, 8
  Drive Filtering setting in, 203
  finishing on, 25
  importance of last frames, 36
  importing and exporting with, 85
  logging in either MediaLog or, 13
  moving inside, 58–61
  object-oriented program, 50
  pre-loading process and, 78
  reference to Avid software, 3
  tape deck connected to, 12
  tape names important to, 110
  ways to do functions on, 18
  workflow involved in finishing on, 127
Media
  backing up, 67–69
  EditCam, 6
  management, 50–52, 179–80, 204
  objects, 67
Media Cleaner Pro, 101–2
Media Composer, advantages over Avid Xpress, 127–28
Media Database, 52–54
Media Relatives, 63–64
Media Tool for editing, 42–43
MediaDocks, 57
MediaFiles folders, 52, 74
MediaLog
  alternatives to, 15
  ships with Avid editing system, 15
  software, 7, 14
  using, 10
MediaShare, 71–72
Memory settings, 180
Messages, error, 163–64

MezzoArchiver, 68–69
Mistakes, fixing digitizing, 61
Mixers, output of tones goes through, 142
MOB (media object) ID, 67
Modems, internal, 7
Modes, sort, 120–21
Modify, using, 9–10
Monitoring, audio, 31–32
Monitors
  adjusting levels, 137–39
  computer, 131
  connections, 165
  consumer-type, 130
  video, 80–81, 131
MOS silent film transfer, 54
Motion video, 96–103
Mouse jog defined, 32–33
msmMac.pmr, 54
msmOMFI.mdb, 54
MUI, 29
Multicam resolutions, 128
MVP solutions, 15

Names
  project, 66–67
  tape, 13–14
NDF (non-drop-frame), 12–13, 207
Nesting 192-96; *See also* Autonesting
Net, Sneaker, 73–74
Networks, 70–71
NewsCutter, 3
Non-timecoded
  material, 45
  tape, 157
Nonlinear video assistants, 201–8
Nonlinearly, navigating, 33
Norton Disk Doctor, 175
NTSC (National Television Standards Committee)
  format, 125
  videotape, 12
Numbers, version, 181–82
NuVista+ video card, 128

Offline cut in online, 108
OMFI (Open Media Framework Interchange), 21, 97, 103–4, 114
  MediaFiles folders, 52, 74
Online, linear, 105–26
Online, offline cut in, 108
Option keys combined with Command keys, 35
Option-add edits, 41
Organizing, 42
Output, 205

Paint and animate, 197–99
PAL (Phase Alternating Line) television system, 14, 125
Photoshop, 96
PICT
  files, 29, 118, 157
  format, 90–91
  sequences, 96, 99–101
PIP (Picture in Picture), 189
Playback
  digitized tones, 142
  reducing delay of, 78–79
PlayLength, 79
Pluges (Picture Line Up Generating Equipment), 134
Ports, SCSI, 44
Ports, serial, 180–81
Power, electrical, 49–50
PowerBook, ADB ports, 11
PRAM (parameter RAM), 180–81
Pre-loading process, 78–79
Precomputes, auto-deletion, 56–57
Prepare, time to, 16
Pro Tools, using OMFI for, 103–4
Problems, multiple methods to solve, 32–33
Process, pre-loading, 78–79
Programs
  disk diagnostic, 175–76
  FileMaker Pro, 15
  shareware screen capture, 29
  third-party EDL, 106
Projects
  backing up, 70
  names, 66–67
  workflow of nonlinear, 5–26
Projects, finishing, 127–61
  adjusting monitor levels, 137–39
    adjusting output levels, 137
    adjusting output levels of video decks, 138
    in Avid Video Tool, 138–39
    color bar signals, 138
    outputting signals, 137
    remote control panel, 139
    several schools of thought, 137
  alpha channels, 153–55
  assembling edits, 155–56
  audio levels, 141–43
    adjusting audio playback controls, 141
    Audio Media II (NuBus) cards, 141–43
    Audio Media III (PCI) cards, 141–43
    Audio Tool, 141
    inputting four channels of audio, 142
    output of tones goes through mixers, 142
    playback digitized tones, 142
  choosing inputs, 131–33
    Betacam components, 132–33
    compatible with EditCam media files, 133
    composite signals, 132
    input ports for S-video, 132
    serial digital ABVB, 133
  color bars and tones on master tapes, 157
  decompose, 144–45
  digitizing scenarios, 145–53
    drive space and offline media, 145–48
    low disk space variation, 149
  equipment, 130–31
    color correctors, 131
    color-measuring devices, 130
    computer monitors, 131
    consumer-type monitors, 130
    vectorscopes, 131
  final cut, 159–60
  importing color bars, 157–59
  importing EDLs, 160
  Media Composer advantage over Avid Xpress, 127–28
  outputting to non-timecoded tape, 157
  picking right AVR, 128–30
  relinking graphics without alpha channels, 153–55
  rendering, 154
  timecodes on master tapes, 154–55
  setting levels, 133–37
    adjustments for video levels, 136–37
    field bars, 135–36
    hue adjustments with composite signals, 136
    memorizing scan lines, 134
    SMPTE bars, 134
    video levels in ABVB, 133–34
  simple redigitizing, 143–44
  studio speakers, 131
  vertical shifting, 139–41
    editing mode, 140
    longitudinal timecode (LTC), 139
    record inhibit switch, 140
    setting correct levels, 140–41
    vertical interval timecode (VITC), 139
  VTR Emulation, 161
Prosumer formats, 11

QuickTime, 97–99, 101
  audio for, 102–3
  Avid codec, 97–98
  importing with Alpha channels, 98–99
  sequences, 96
  size, 98

RAM (random access memory)
  buying more, 51
  Caps Lock keys take up lot of, 31
  how much in system, 78
  never enough, 173

Record inhibit switch, 140
Redigitizing, 143–44, 206
Reels, dupe, 121–24
Relinking, 74–76, 153–55
Remote control panel, 139
Render on the Fly, 80
Rendering, 23–24
  effects, 188–89
  for speed, 81
Resolutions
  linked closely to size, 90
  multicam, 128
Reveal File, 64
Rewind, and fast forward, 33
RGB vs. CCIR for import, 92–94
Room design, 47–48
RT-11 format, 112–14
RTFM (Read The Manual — Please), 163


S-video, input ports for, 132
Scan lines, 134
Scrub, audio, 31
Scrubbing and Caps lock keys, 31
SCSI (Small Computer System Interface)
  cables, 47
  connections, 166–70
  MediaShare, 72
  ports, 44
Sequences
  consolidating, 59
  PICT, 99–101
  QuickTime or PICT, 96
Serial digital ABVB, 133
Serial ports, 180–81
Servers, 73
Sessions, between, 57–58
Settings, user, 70
Shareware screen capture program, 29
Signals
  color bar, 138
  composite, 132
SKII (sound designer) format, 21
Sliding and slipping, 40
SMPTE (Society of Motion Pictures and
    Television Engineers)
  bars, 134
  wipe codes, 116
Sneaker Net, 73–74
Software
  accessing to original, 172
  backup, 44
  Compact Pro, 44
  EDL, 110
  isolating hardware from, 176–77
  MediaLog, 7
  problems, 177

Sony
  and DOS, 114
  editing decks, 108
Sort modes, 120–21
Sound; *See also* Audio
  levels, 118–19
  preparing, 107–8
  tracks, 21
Sound mixing
  challenges of, 21
  and color corrections, 20–23
Source table, 124–25
Space, empty, 57
Spatial compression defined, 102
Speakers, studio, 131
Speed, and keyframes, 191
Storage estimates, 54–56
Storage Tool, 16
StorageCalc, 16
Studio speakers, 131
Support, customer, 183–84
Switch, record inhibit, 140
Sync locks, 38–39
System performance, improving, 77–83
  building pipes, 78–79
  minimizing timeline redraw time, 81–83
  moving video monitor, 80–81
  playLength, 79
  pre-loading process, 78–79
  Render on the Fly, 80
  rendering for speed, 81
  video mixdown for speed, 81


Tables
  Fixed Q, 55–56
  source, 124–25
Tablets, graphics, 91
Tape names
  in EDLs, 110
  importance of, 13–14
Tapes
  blacking, 206–8
  master, 154–55, 157
  non-timecoded, 157
  wear and dubbing, 10–11
TBC (time base correctors), 22, 121, 138
Techniques, intermediate, 27–45
  audio monitoring, 31–32
  backing up, 43–44
  fast forward and rewind, 33
  importance of last frames, 36
  jumping precisely, 34–35
  Media Tool for editing, 42–43
  miscellaneous modifiers, 35
  multiple methods to solve problems, 32–33
  navigating nonlinearly, 33

non-timecoded material, 45
organizing, 42
other hand, 32
speeding up dragging, 35
trimming, 36–41
  mastering trim, 37–38
  option-add edits, 41
  slipping and sliding, 40
  staying in sync, 38
  sync locks, 38–39
  trim vs. extend, 37
  trimming in two directions, 39
using keyboards, 27–31
  custom keyboards, 28–29
  J-K-L keys, 29–30
  J-K-L for trimming, 30
  scrubbing and Caps lock keys, 31
using timelines, 33–34
Television graphics, 91–92
Templates, effect, 191
Temporal compression defined, 102
Termination defined, 168
Terran Interactive, 101
3-D effects, 196–97
Time to prepare, 16
Timecodes; *See also* Non-timecoded material;
    Non-timecoded tape
  breaks, 7–9
  burn-in, 13
  consistency of, 14
  decomposing without, 150–51
  dubbing with, 108–10
  increments continuously, 109
  keeping from original source, 109
  laying down new, 109
  on master tapes, 154–55
  time-of-day (TOD), 7–8
  using, 110
Timeline User Settings, 83
Timelines
  minimize redraw times, 81–83
  using, 33–34
Titles
  recreating, 151–52
  uncompressed, 199–200
TOD (time-of-day) timecode, 7–8
Tones, playback digitized, 142
Transcoders defined, 133
Trimming, 36–41
  J-K-L for, 30
  in two directions, 39
Troubleshooting, 163–84
  Apple Desktop Bus (ADB) cables, 166
  audio connections, 170
  audio distortion, 178
  calling customer support, 183–84

computer woes, 171–74
  accessing to hardware, 173
  accessing to original software, 172
  extension conflicts, 171–72
  never enough RAM, 173
  starting up from another drive, 173–74
  working from floppies, 172
electrostatic discharge (ESD), 182–83
error messages, 163–64
fonts, 178–79
importance of connections, 164–65
isolating hardware from software, 176–77
maintenance, 174–76
  disk diagnostic programs, 175–76
  rebuilding desktops, 174–75
media management, 179–80
monitor connections, 165
serial ports, 180–81
Small Computer System Interface (SCSI)
    connections, 166–70
  connecting wide and narrow drives, 169–70
  ID, 168–70
  rules, 167–68
  termination, 168
software problems, 177
version numbers, 181–82

Unity defined, 137
Unlink, link broken using, 76
Unlinking of master clip, 74
UPS (uninterruptible power source), 49–50
User settings, 70

Vectorscopes, 131
Version numbers, 181–82
Vertical shifting, 139–41
Video
  adjusting levels, 136–37
  adjusting output levels of decks, 138
  cleaning heads, 206
  and graphics, 117
  mixdown for speed, 81
  monitors, 80–81, 131
  motion, 96–103
  NuVista+ cards, 128
Video assistants, nonlinear, 201–8
  backing up, 205
  blacking tape, 206–8
  digitizing, 202–3
  drives, 203–4
  maintenance, 204–5
  media management, 204
  output, 205
  redigitizing, 206

Videotapes
  digital, 108
  NTSC, 12
VITC (vertical interval timecode), 11–12, 139
VTR Emulation, 161

Wipe codes, Society of Motion Pictures and Television Engineers (SMPTE), 116
Workflow of nonlinear project, 25–26
  change one's mind, 5–6
  changes easy with nonlinear systems, 6
  creating with one's own eyes, 19
  drop-frame (DF) vs. non-drop-frame (NDF), 12–13
  editing, 17–19, 18–19
  finishing, 24
  first nonlinear assembly, 19
  input, 6–17
    alternatives to MediaLog, 15
    batch digitizing, 16–17
    creating databases, 15
    importance of tape names, 13–14
    logging, 6–7
    logging by hand, 11–12
    organizing bins, 15–16
    tape wear and dubbing, 10–11
    timecode breaks, 7–9
    using Modify, 9–10
  keeping audio, 24–25
  planning for changes, 25–26
  refining what one has done, 18–19
  rendering, 23–24
  sound mixing and color corrections, 20–23
  soundbites and concern about graphics, 19
  workflow outside suite, 20
  working with graphics, 20
Workstations, graphics, 91–92

Zip or Jaz drives, 44

## ABOUT THE AUTHOR

Steve Bayes has been designing and organizing editor workshops and seminars around the world since 1993. He has recently served as Senior Instructor at Avid Technology, Inc, where he has assisted in the development of the courseware being used by the Avid Authorized Education Centers and by the 100 Avid Certified Instructors for Media Composer and MCX. As an editor, postproducer and consultant Steve has worked extensively in advertising, educational and long format television and video, and has been the recipient of numerous awards. He is currently a designer of the Avid Media Composer.

778.593/BAY
S.L.